Joseph Lerner

ANXIETY

BY DONALD W. GOODWIN

Anxiety

The Writer's Vice

Psychiatric Diagnosis, third edition,
Co-authored with S. B. Guze

Phobia: The Facts

Longitudinal Research in Alcoholism
*Co-authored with K. T. Van Dusen
and S. A. Mednick*

Alcoholism: The Facts

Alcoholism and Affective Disorders
Co-edited with Carlton Erickson

Is Alcoholism Hereditary?

ANXIETY

Donald W. Goodwin

New York   Oxford
OXFORD UNIVERSITY PRESS
1986

Oxford University Press

Oxford New York Toronto
Delhi Bombay Calcutta Madras Karachi
Kuala Lumpur Singapore Hong Kong Tokyo
Nairobi Dar es Salaam Cape Town
Melbourne Auckland

and associated companies in
Beirut Berlin Ibadan Mexico City Nicosia

Copyright © 1986 by Oxford University Press, Inc.

Published by Oxford University Press, Inc., 200 Madison Avenue
New York, New York 10016

All rights reserved. No part of this publication may be reproduced,
stored in a retrieval system, or transmitted, in any form or by any means,
electronic, mechanical, photocopying, recording, or otherwise,
without the prior permission of Oxford University Press.

Library of Congress Cataloging in Publication Data
Goodwin, Donald W.
Anxiety.
Bibliography: p. Includes index.
1. Anxiety. I. Title.
RC531.G66 1986 616.85′223 85-7179
ISBN 0-19-503665-4

Printing (last digit): 9 8 7 6 5 4 3 2 1
Printed in the United States of America

To Georgia

Preface

In 1950 Rollo May published a book called *The Meaning of Anxiety*. He wrote it while convalescing from tuberculosis, an anxious occasion. He mentioned knowing of only two other books devoted to anxiety—one by Kierkegaard, the other by Freud.

By the time May revised the book in 1977, the literature on anxiety had grown enormously. Philosophers and theologians located anxiety at the heart of human experience. Existentialists like Sartre, Camus, Tillich, and May himself were as widely sold on the college campuses during the 1950s and 1960s as they were mostly unread and sometimes unreadable. Today, their popularity seems in decline, but meanwhile the social scientists had discovered anxiety and for a time words like anomie and alienation, roughly synonymous with anxiety, had a vogue on the campuses (now, also, seemingly in decline).

Other trends emerged. In the 1950s the pharmaceutical industry discovered that anxiety-relief was profitable (as the brewers and distillers had always known). Simultaneously, the Pavlovians and Skinnerians were finding new ways to make animals anxious, or at least deathly afraid. The last group to inquire into the origins of anxiety were the chemists, who, commanding a stunning new technology undreamed of in the 1950s, found switches for anxiety in the brain that could be turned off and turned on.

It is time to pull together what all these people have been doing and determine where we now stand. The time is particularly ripe because of four recent developments:

1. The invention and proliferation of synthetic chemicals that specifically relieve anxiety.
2. New knowledge about brain chemistry and anxiety.
3. Development of a new diagnostic scheme that divides pathological forms of anxiety into separate disorders.
4. New treatments for the specific anxiety disorders.

Chapter 1 makes the important distinction between anxiety and fear. Anxiety is diffuse, vague—an "attack from the rear," to use Kurt Goldstein's phrase. Fear is a response to a concrete, knowable danger. This book holds that fear is sometimes useful but that anxiety—as defined here—never is.

The next two chapters review what philosophers and biologists have written about anxiety, beginning with the existentialists. The physical aspects of anxiety are presented, starting with the simple startle reaction, followed by a description of those parts of the nervous system that make anxiety "happen." Some fears we are born with and others we learn and these are discussed. The psychologists' views of anxiety follow—especially those of Freud and Watson, whose radically divergent ideas on the subject dominated twentieth-century psychology.

The six chapters in Part I represent what "came before." What's happening now—and what happens next—is the subject of Part II. It begins with a Polish chemist working for a Swiss company in a New Jersey laboratory who almost threw away a chemical that has revolutionized the treatment of anxiety. It reviews important new developments in the neurosciences and asks the question, "Does the brain make its own drugs?" Finally, it describes how the psychoanalytic movement in America is gradually being replaced by a more pragmatic approach to mental illness.

Part III deals with anxiety disorders. The American Psychiatric Association identifies eight such disorders and each is described in some detail. The purpose of this book is not self-diagnosis, much less self-treatment, but the reader with an anxiety disorder can at least form a reasonable idea of what it might be. A recent study by the National Institute of Mental Health found that more than 13 million Americans suffer from these disorders.

For scholars, professionals, and those who are just curious, the book ends with sundry comments and sources of information. The book is about women as well as men but individuals are usually referred to in the male gender since this remains a convention in books about both sexes.

Kansas City   D.W.G.
July 1985

Acknowledgments

My thanks to Dr. Elizabeth Penick and Dr. Ronald Martin, and my wife, Sally, for reading the manuscript and making many helpful suggestions. I am also grateful to Professor Robert Hudson, Chairman of the Department of The History of Medicine at the University of Kansas, for contributing points of historical interest. The secretarial talents of Mrs. Evelyne Karson were, as always, simply invaluable.

Finally, my thanks to Jeffrey House of Oxford University Press, whose friendship and editorial gifts have made a dozen years of writing books for Oxford as enjoyable as such enterprises can ever be.

Some of the information in this book appeared in substantially altered form in *Phobia: The Facts* (Oxford University Press), published originally in England in 1983.

Contents

I. THE FEAR WITHOUT A NAME

1. A Question of Definition — 3
2. The Discovery of Dread — 7
3. The Physiology of Fear — 14
4. Innate Fears — 23
5. The Role of Conditioning — 32
6. Psychological Theories — 40

II. A NEW ERA

7. The Invention of Antianxiety Drugs — 51
8. Does the Brain Make its own Drugs? — 63
9. The New Classification — 79

III. THE ANXIETY DISORDERS

10. Rules — 97
11. Generalized Anxiety Disorder — 101
12. Panic Disorder — 112
13. Phobia: An Introduction — 123
14. Simple Phobias — 128
15. Social Phobias — 138
16. Agoraphobia — 145
17. Separation Anxiety Disorder (School Phobia) — 152
18. Treatment for Phobias — 157
19. Obsessive Compulsive Disorder — 171
20. Post-Traumatic Stress Disorder — 178

Notes and References — 191

Index — 227

PART I

THE FEAR WITHOUT A NAME

A continual Anxiety for Life vitiates all the Relishes of it, and casts a Gloom over the whole Face of Nature; as it is impossible we should take Delight in any thing that we are every Moment afraid of losing.

Joseph Addison
The Spectator

As the order of existence becomes increasingly rational and is applied to everyone, its extraordinary success carries with it a feeling of impending doom and of an anxiety which develops because it is no longer apparent what makes life worthwhile.

Karl Jaspers
Die geistige Situation der Zeit

1

A Question of Definition

> Such is man that if he has the name for something, it ceases to be a riddle.
>
> Isaac Singer

"For the past 100 years," Rollo May wrote in 1950, "students of humanity have been increasingly preoccupied with a nameless and formless uneasiness that has dogged the footsteps of modern man." He was talking about anxiety. Not much was known about anxiety when he wrote those words. Today, more is known. People are even talking about doing something about anxiety.

Doing something about anxiety? Just saying it sounds wrong. Isn't anxiety *normal*? Something everyone has? Lord knows, isn't there plenty to be anxious about? Doing something "about" anxiety sounds as futile—as counterproductive—as doing something about the sex drive or the weather.

Much depends on how one defines anxiety. This book is based on the premise that *nobody* needs or wants anxiety. As the word is defined here, anxiety is an emotion that signifies the presence of a danger that cannot be identified, or, if identified, is not sufficiently threatening to justify the intensity of the emotion. Anxiety is an undesirable emotion. It leads to nothing useful because the true source of the distress is unknown. Anxiety is like an itch that can't be scratched because it is about two inches left of your third thoracic vertebra—no, down a little, up slightly . . . *yes*, no, . . . damn!

Anxiety is different from fear. Fear signifies the presence of a *known* danger. The strength of fear is more or less proportionate to the degree of danger. In general, fear is a desirable emotion because it leads to much that is useful—avoiding danger, for example.

Anxious people often *think* they know what they are afraid of: the garden snake, the crowd in the theater, a rebuke from the boss. But, if they think about it, they recognize that the anxiety is disproportionate to the threat. Thus, the real source of the distress is unknown, even if some things seem to bring it on more than others.

Not everyone agrees with this definition. Most psychiatrists, starting with Freud, distinguish anxiety from fear more or less as above. But many think that some anxiety is necessary, even good for you. It builds character, enhances creativity, enlarges awareness of life's possibilities. Actors act better, they say, if they feel a little anxious. Anxiety keeps us on our social toes. Criminals commit crimes because they *don't* have anxiety. Hurray for anxiety!

Hurray, anyway, for "normal" anxiety. Everyone recognizes that abnormal anxiety is not good. When a woman panics in the grocery store line, this is abnormal. When she feels anxious about her husband's drinking, this is normal.

The problem is semantic and occurs when the words fear and anxiety are used interchangeably. As the terms are defined in this book, the woman is *anxious* in the grocery story, but *afraid* of her husband's drinking. The consequences of the latter are knowable and specific, the fear is appropriate and may lead to useful action, like divorce. There is nothing useful about panicking in a grocery store; no appropriate action is possible; the source of the distress is unknown (truly unknown, despite the theories of Freud and others).

Rollo May, in his 1950 book, *The Meaning of Anxiety*, gave an example of "normal" anxiety:

> A prominent Socialist was living in Germany when Hitler came into power. Over a period of some months he knew that some of his colleagues were being imprisoned in concentration camps or taken off to other unknown fates. During this period he lived in the perpetual awareness that he himself was in danger, but he never could be certain *if* he would be apprehended, or, if he were, *when* the Gestapo would come, or, finally, *what* would happen to him if he were arrested. Throughout this period he experienced the diffuse, painful, and persistent feelings of uncertainty and helplessness which are characteristic of anxiety . . . yet it was proportionate to the actual threat and could not be termed neurotic.

Two points need to be made. First, May says that uncertainty and helplessness are characteristic of anxiety. They also are characteristic of fear. A man in the jungle worried about a tiger jumping out from be-

A QUESTION OF DEFINITION 5

hind a tree is uncertain about when or if the tiger will jump and he also feels pretty helpless. But he knows what concerns him: the tiger. The German Socialist was afraid, not anxious (as I use these words).

Second, May says the Socialist's anxiety was not neurotic. What does this mean? Neurotic anxiety is a concept introduced by Freud. It refers to a feeling of danger arising from unconscious conflicts. Being unconscious, the nature and even existence of the conflicts is hidden, but, for Freudians, these conflicts involve sexual and aggressive feelings directed toward the parents in childhood.

Freud stressed that the danger is real, not imaginary—at least for the child. Making sexual advances toward one's mother and physically attacking one's father can indeed have terrible consequences for a child (the most terrible and commonly imagined one being castration, according to Freud). Since the conflict is repressed because the consequences are too terrible to entertain in conscious thought, the person never has an opportunity to grow out of the conflict. All he can do is endure the signals of anxiety sent up from the long-repressed conflict and try to defend himself.

In defending himself, he constructs psychological defenses calculated to reduce the anxiety. Examples are *phobias* (whereby formless anxiety is shaped into specific fears) and *sublimation* (whereby the anxiety leads to works of art or good deeds and is useful to both the individual and society). If harmful, the defenses are called neurotic; if useful, they are regarded as healthy. Anxiety that cannot be transformed into neurotic symptoms or useful activity is experienced as "free floating" and characterizes those people described later in the book as having generalized anxiety disorder.

Later psychoanalysts did not confine the roots of neurotic anxiety to unconscious conflicts involving sex and aggression toward the parents. Anxiety, they said, could arise from conflicts involving dependency needs (Horney), security needs (Sullivan) and a need for power (Adler). But the theories all share a common element: anxiety indicates the presence of a problem that needs to be solved but cannot be solved because the person who has the problem is unaware of its nature. Bringing the problem into conscious awareness is the goal of most forms of psychoanalysis. This is supposed to relieve the anxiety and the symptoms it produces. Whether it actually does so is open to question.

The unconscious conflict theory of neurotic anxiety has come to be accepted by large numbers of therapists, intellectuals, and the educated public. The theory may indeed be true, but there is hardly any evidence of a scientific nature to support it. It would be more open-minded to view anxiety as a "nameless and formless uneasiness" (to use May's words again), the cause of which remains unknown.

Still, not knowing the cause of something is intolerable for many people. They demand explanations. They want names for things and, once given a name, the name becomes the explanation.

Philosophers, biologists, and psychologists are proficient at naming things. What they have to say about anxiety is interesting and perhaps useful and, until quite recently, it was the only information we had on the subject. The next three chapters deal with anxiety as it has been described by philosophers, biologists, and psychologists.

2

The Discovery of Dread

> And no Grand Inquisitor has in readiness such terrible tortures as has anxiety, and no spy knows how to attack more artfully the man he suspects, choosing the instant when he is weakest, nor knows how to lay traps where he will be caught and ensnared, as anxiety knows how, and no sharpwitted judge knows how to interrogate, to examine the accused, as anxiety does, which never lets him escape, neither by diversion nor by noise, neither at work nor at play, neither by day nor by night.
>
> Sören Kierkegaard
> *The Concept of Dread*

Sören Kierkegaard discovered anxiety in 1844. He announced his discovery in a little book called *The Concept of Dread*. Translated into English 100 years later, it is a classic of modern existentialism.

Obviously anxiety existed before Kierkegaard, and indeed anxiety disorders have been known throughout history. The Hippocratic writings describe at least two phobic persons. One was "beset by terror" whenever he heard a flute, while the other could not go beside "even the shallowest ditch" and yet could walk in the ditch itself. Obsessional traits were recorded many centuries ago. In the early seventeenth century Richard Flecknoe described an "irresolute Person who hovers in his choice, like an empty Ballance with no waight of Judgement to incline him to either scale." In the same century Robert Burton even distinguished "morbid" fears from "normal" fears in his *Anatomy of Melancholy*. Demosthenes' stage fright was normal; Caesar's fear of sitting in the dark was morbid.

Nevertheless, Kierkegaard is credited with the first description of anxiety as a vague, diffuse uneasiness, different from fear in that no apparent danger is present, and pervasive, allowing no escape, "neither by diversion nor by noise, neither at work nor at play, neither by day nor by night."

Kierkegaard was the founder of existentialism. A few words about the man and the philosophy are essential to any book on anxiety.

KIERKEGAARD WAS BORN in Denmark in 1813, the son of a wool merchant. As a young man his father cursed God, and Sören spent much of his life trying to decide whether to follow his father's example. He saw life as a case of *Either/Or* (the title of his best-known book): either you believed in Christ and led a devout life or rejected Christ and led a life of pleasure. It really didn't matter which, Kierkegaard said. You would still feel anxious. Anxiety was intrinsic to the human condition: a fundamental tenet of existentialism.

Believing in God, Kierkegaard said, required a "leap" of faith. There was no proof of God's existence; seeking proof itself was irreligious. But one had the *freedom* to choose. It was this freedom, without any guarantees about the correct choice, that caused anxiety. Reading Kierkegaard reminds one of the executive's famous lament: "Decisions, decisions, decisions."

THERE ARE TWO kind of existentialists—theists and atheists—but the ones who believe in God view the world almost exactly as the ones who do not. Whether one "chooses" to believe or not believe, the world is a miserable place. Pascal, a seventeenth-century predecessor of Kierkegaard, described "existence" as well as any existentialist:

> When I consider the brief span of my life, swallowed up in the eternity before and behind it, the small space that I fill, or even see, engulfed in the infinite immensity of spaces which I know not, and which know not me, I am afraid, and wonder to see myself here rather than there; for there is no reason why I should be here rather than there, now rather than then.
>
> On beholding the blindness and misery of man, on seeing all the universe dumb, and man without light, left to himself, as it were, astray in this corner of the universe, knowing not who has set him here, what he is here for, or what will become of him when he dies, incapable of all knowledge, I begin to be afraid, as a man who has been carried while asleep to a fearful desert island, and who will wake not knowing where he is and without any means of quitting the island. And thus I marvel that people are not seized with despair at such a miserable condition.

THE DISCOVERY OF DREAD 9

Kierkegaard *was* seized by despair, but not from the miserable condition. Kierkegaard's despair came from his awareness of good and evil and the burden of personal choice. Twentieth-century existentialists of the atheist persuasion—Jean Paul Sartre being the most famous—avoided words like good and evil, but also emphasized the awesome responsibility that came from the freedom to make choices.

For Sartre and his followers, life has no meaning or purpose, but individuals can "will" meaning into their existence by making choices that *affirm* themselves as individuals. The first and most important choice is whether to commit suicide. Once suicide is rejected, then life becomes one choice after another, with each choice, ultimately, regardless of the language, a choice between good and evil.

The good choice—for the twentieth-century existentialist—is the courageous choice, the selfless choice, the choice to act *as if* life had purpose or meaning. Become committed, Sartre said: join the Resistance, write a book, run for office. Become committed even when, like Sisyphus, you know your exertions are in vain. There will be no reward in heaven and probably none on earth.

The human situation, as Sartre saw it, was absurd and tragic, but humans could still achieve integrity, nobility, and valor. Even in failure man could defy the world. He could exist, create a "self."

THE EXISTENTIALISTS agree that it is the freedom to choose, not the Godless world, that causes dread and anxiety. "When I behold my possibilities," Kierkegaard wrote, "I experience that dread which is the dizziness of freedom, and my choice is made in fear and trembling." The more choices—the more awareness of choice—the more anxiety. Creative people are especially anxious because creativity involves many choices. Ordinary people try to escape choices. Confronted with difficult situations, they say "It's impossible. Nothing can be done." They lose themselves in sports, entertainment, sex, hobbies, drink, anything to avoid the awareness of freedom. Kierkegaard spoke of the nineteenth century as a "cowardly age" in which "one does everything possible by way of diversions and the Janizary music of loud-voiced enterprises to keep lonely thoughts away, just as in the forests of America they keep away wild beasts by torches, by yells, by the sound of cymbals." (Today, he probably would have substituted rock concerts.)

But attempts to evade anxiety are doomed to failure. In running from anxiety, he said, a person loses his most precious opportunity for the emergence of one's self and for one's education as a human being. "If a man were a beast or an angel, he would not be able to be in anxiety. Since he is both beast and angel he can be in anxiety and the greater the anxiety the greater the man." Kierkegaar regarded anxiety as a "school which teaches people to face death and accept a human situation frankly. . . . When a person goes out from the school of possibility and knows, more thoroughly than a child knows the alphabet, that he can demand of life absolutely nothing, and that terror, perdition, annihilation, dwell next door to every man, and has learned the profitable lesson that every anxiety which alarms may the next minute become a fact . . . he will extol reality."

Those who refuse thus to extol reality deserve no pity. Sounding like a true existentialist, the philosopher Walter Kaufmann wrote: "All man's alibis are unacceptable: no Gods are responsible for this condition; no original sin; no heredity and no environment; no race, no caste, no father and no mother; no wrong-headed education, no governess, no teacher; not even an impulse or a disposition, a complex or a childhood trauma. Man is free; but his freedom does not look like the glorious liberty of the Enlightenment; it is no longer the gift of God. Once again, man stands alone in the universe, responsible for his condition, likely to remain at a lowly state, but free to reach above the stars."

Kierkegaard chose to be a Christian. Sartre chose to be an atheist. But the important thing is they *chose*. This is the message of existentialism.

Existentialism is not a philosophy but, as Kaufmann said, a label for several widely different revolts against traditional philosophy. Most of the living "existentialists" have repudiated the label. A bewildered outsider, Kaufmann said, might conclude that the only thing they have in common is a marked aversion for each other.

Existentialism offers nothing resembling a metaphysical system or ethical doctrine. The three writers who appear on every list of existentialists—Jaspers, Heidegger, and Sartre—seem to agree on only one thing: man is free to make choices and that freedom is the source of his anxiety.

Existentialism is new, a post-World War II phenomenon. Bertrand Russell, in his 1945 monumental history of philosophy, mentioned neither Kierkegaard nor existentialism. Kaufman calls it a "timeless sensibility that can be discerned here and there in the past; but it is only in recent times that it is hardened into a sustained protest and preoccupation."

Why did this hodgepodge of gloomy ideas arise at this moment in history? There are several reasons, all bearing on why anxiety has also become a preoccupation of our times.

In Western civilization the Age of Faith gave way in the seventeenth and eighteenth centuries to the Enlightenment, or Age of Reason. In the Middle Ages, faith in religion, the role of kings, and the medieval social order provided a rigid set of instructions for personal behavior. Doubt—the father of anxiety—was uncommon. People in the Middle Ages knew who they were and what was expected of them. While often afraid, presumably they were less often anxious (if doubt indeed is the father of anxiety).

When the emphasis changed from Faith to Reason, this opened the door to doubt and thus to anxiety. At first, however, faith in Reason was as absolute as faith in God. Spinoza, a seventeenth-century rationalist, believed that fear was inseparable from hope and that both were weaknesses easily overcome by reason. He believed serenity could be achieved by simply "ordering our thoughts and images"—and Spinoza, by all accounts, was a serene man. Hegal wrote books demonstrating that nothing was more reasonable than the existence of God. The eighteenth century was a time of great optimism and belief in progress, when science, democracy, and reason would surely lead to ever-increasing prosperity, health, and happiness.

It didn't happen that way. To be sure, Europeans and Americans in the twentieth century are incomparably more healthy than their medieval counterparts. But it is questionable whether they are happier, and prosperity has eluded large parts of the globe. Industrialization, urbanization, material comforts, mobility: none has seemed to promote serenity.

By the mid-nineteenth century, Kierkegaard's time, the power of reason had come under severe attack. Passion, not reason, was the rallying cry of Rousseau and the Romantic movement. The death knell of Reason, so to speak, came in the French Revolution when cathedrals were converted to Temples of Reason. Consider the following, from a recent book on the Revolution:

> . . . a few days before the Commune ordered the closure of all churches in the city, a grand Festival of Reason was celebrated in Notre Dame. A

young actress was carried into the cathedral by four citizens to represent the Goddess of Reason. Clothed in white drapery with a blue cloak over her shoulders and a red cap of liberty crowning her long hair, she was accompanied by a troupe of girls also dressed in white with roses on their heads. She sat on an ivy-covered chair while speeches were made, songs were sung, and soldiers paraded about the aisles carrying busts of martyrs of the Revolution.

This occurred while wholesale butchery was turning the streets of Paris into rivers of blood.

Existentialism was a revolt against Reason. "Reason is a whore," Kierkegaard said, easily bought. Faith in Reason went the way of religious faith, discredited as, at best, a frail reed.

When man lost faith in religion and reason, what *could* he believe in? Nothing, the existentialists said. It wasn't important that he believed in anything. He could still choose and act. But of course he would be anxious—very anxious. So anxious, indeed, that in 1950, influenced by the existentialists, W. H. Auden wrote a long poem called *The Age of Anxiety*, and the term caught on. The Age of Anxiety has been attributed to loss of religious faith, a decline in morals, the deterioration of family and community life, disillusionment in science, excessive individualism, materialism, secularization, industrialization, urbanization, the threat of nuclear annihilation—all leading to inner confusion, alienation, uncertainty with respect to values and acceptable standards of conduct, leaving us, in Auden's words, as "unattached as tumbleweed."

To this formidable list may be added another reason for the Age of Anxiety: for many, there is less to fear.

A splendid cure for anxiety is being chased by a lion: fear takes precedence over anxiety. People today who live in the affluent, industrialized, welfare-state democracies—meaning primarily countries in North American and Western Europe—are safer and more secure than peoples living at any other time in history. It is hard for modern man in these countries to realize just how *frightening* life used to be.

Here is how life for the average eighteenth century Frenchman was described by Robert Darton:

> During the last few decades research has uncovered a Malthusian society, in which the basic fact of life was the inexorable struggle against death. Most Frenchmen lived in or near a state of chronic malnutrition and offered little resistance when plague and famine sliced through the popula-

tion. For every ten babies born, two or three died before their first birthday and four or five died before the age of 10. Marriages usually lasted only about 15 years, terminated by death rather than divorce. [France] was a society of widows and orphans, of stepmothers and Cinderellas. A bad harvest could drive hundreds of thousands below the line that separated poverty from indigence, forcing them on the road, where they lived by their wits, begging and stealing, until in the end they surrendered in a pestilential poorhouse or crawled under a bush or a hayloft and died—*croquants* who had "croaked."

The human condition has changed so drastically over the last two centuries that we can barely imagine the way it appeared to people whose lives really were nasty, brutish, and short.

Darton went on to say that when dreams came true in French fairy tales, they left the hero in the barnyard rather than in never-never land. "It was a well-stocked barnyard and the hero had just enjoyed a good meal. Wish fulfillment usually took the form of food. . . . Once supplied with magic wands, rings, or supernatural helpers, the first thought of the peasant hero was always for food. He never showed any imagination in his ordering. He merely took the *plat du jour* and it was always the same: solid peasant fare. . . . Usually the peasant raconteur [telling the fairy tale] does not describe the food in detail. Lacking any notion of gastronomy, he simply loads up his hero's plate; and if he wants to supply an extravagant touch, he adds, 'There were even napkins.' "

Where there was little food, there was not much notion of gastronomy. Where there was much fear, anxiety was not a problem. The Age of Anxiety could also be called the Age of the Fortunate.

3

The Physiology of Fear

> These changes—the more rapid pulse, the deeper breathing, the increase of sugar in the blood, the secretion from the adrenal glands—were very diverse and seemed unrelated. Then, one wakeful night, after a considerable collection of these changes had been disclosed, the idea flashed through my mind, that they could be nicely integrated if conceived as bodily preparations for supreme effort in flight or in fighting.
>
> Walter B. Cannon
> "The Role of Hunches" in
> *The Way of an Investigator* (1945)

Walter B. Cannon, the great Harvard physiologist, believed in hunches. In his half-century at Harvard, most of it spent as head of the Physiological Laboratory, he came up with many hunches bearing on the influences of emotion on the body and vice versa. His book, *Bodily Changes in Pain, Hunger, Fear and Rage*, was the first systematic account of experimental observations linking emotions to bodily changes.

Cannon was a scientist. His hunches were subjected to controlled tests in the laboratory and if they lacked support, he dropped them. Cannon and Freud lived and worked during the same half-century, but they seem almost unaware of each other's existence. Each had a different modus operandi—Cannon was an experimentalist, Freud a theoretician.

Cannon's belief in hunches seems to resemble theories of the unconscious mind, but Cannon rejected the similarity. Since the existence of an unconscious mind has widely come to be regarded as self-evident, Cannon's reservations are worth quoting at some length:

> There has been much discussion of what lies back of the experience of having hunches. They have been ascribed to the operations of the "subconscious mind." This expression seems to me to be a confusion of terms, for it involves the concept that a mind exists of which we are not conscious. I am

aware that in psychology this view has been held [a sublime understatement].

Indeed, one psychologist with whom I discussed the matter declared that wherever nerves coordinate the activity of muscles, a mind is present. I told him that the nerve net in the wall of the intestine brings about a contraction of muscles above a stimulated point and a relaxation below it so that a mass within the tract is moved onward. This is coordinated action, and I asked him whether he would ascribe a mind to the intestine. His reply was, "Undoubtedly." The attitude thus expressed was extreme. It may be taken, however, as a basis for criticizing the assumption that there is a mind wherever nervous activity goes on, when in fact there is no evidence to support the notion. Numerous highly complex responses which can be evoked from the spinal cord and many nice adjustments made by the part of the brain that manages our normal balance and posture are wholly unconscious. *There is no indication whatever that anything which we recognize as a mind is associated with these nervous activities* [emphasis added].

To me as a physiologist, mind and consciousness seem to be equivalent, and the evidence appears to be strong that mind or consciousness is associated with a limited but shifting area of integrated activity in the cortex of the brain. The physiologist assumes that, underlying the awareness of events as it shifts from moment to moment, there are correlated processes in the enormously complicated mesh of nervous connections in the thin cortical layer. Such activities could go on, however, in other parts of the cortex and at the time be unrelated to the conscious states. They would be similar in character to the activities associated with consciousness, but would be *extraconscious*.

Our knowledge of the association between mental states and nervous impulses in the brain is still so meager that we often resort to analogy to illustrate our meaning. The operation going on in an industry under the immediate supervision of the director is like the cerebral processes to which we pay attention; but meanwhile in other parts of the industrial plant important work is proceeding which the director at the moment does not see. Thus also with extraconscious process. By using the term "extraconscious processes" to define unrecognized operations which occur during attention to urgent affairs or during sleep, the notion of a subconscious mind or subconsciousness can be avoided.

Cannon's advice to avoid concepts like "the unconscious" went largely unheeded. Psychoanalysis, which came to dominate American psychiatry during Cannon's time, was committed totally to acceptance of an unconscious mind. The discipline of psychosomatic medicine was strongly influenced by psychoanalysis and adopted the concept of unconscious mind more or less uncritically. Cannon's skepticism and the careful experimental work carried out by physiologists were largely lost in a tumult of theory. The work provides, nevertheless, much of the hard evidence that exists regarding the connection between emotions and the body. Those aspects relevant to anxiety will be examined here.

A *note on language:* physiologists generally use the word fear instead of anxiety, but in this case the words can be used interchangeably. Anxiety *is* fear, however vague and formless. The physical correlates of fear are also those of anxiety, or at least severe anxiety. Milder forms of anxiety may involve other physical reactions (see Tom's story, a few pages later).

The Startle Reaction

Physiologists and psychologists have long noted similarities between the relatively simple "startle reaction" and the more complex phenomena of fear and anxiety. This, then, is the logical subject with which to begin a discussion of the physical aspects of fear.

If you fire a gun behind a person's head, he will quickly jerk his head forward, blink his eyes, and in other ways exhibit a startle reaction. The response is involuntary and occurs throughout life. It has been much studied in infants and many believe it is the precursor of the emotions of fear and anxiety.

The startle reaction has been best described by two psychologists named C. Landis and W. A. Hunt. Its most dominant feature is a general flexion of the body "which resembles a protection contraction or 'shrinking' of the individual." It involves a blinking of the eyes as well as the following: head movement forward, a characteristic facial expression, raising and drawing forward of the shoulders. Outward movement of the upper arms, bending the elbows, turning the hands backwards or downwards, flexion of the fingers, forward movement of the trunk, contraction of the abdomen, and bending of the knees. "It is a basic reaction, not amenable to voluntary control, is universal, and occurs in infants as well as adults, in the primates and in certain of the lower animal forms." Neurologically, the startle reaction involves an inhibition of the higher brain centers. These centers are unable to integrate a stimulus of such suddenness; we startle before we know what threatens us.

Landis and Hunt describe the startle reaction as "pre-emotional." It is a rapid, transitory response much simpler in its organization and expression than the so-called emotions. Landis and Hunt say emotion may *follow* the startle reaction. After being startled, adults show secondary behavior (emotion) such as curiosity, annoyance, and fear. The authors suggest that this secondary behavior is a "bridge from the innate and unlearned response over to the learned, socially conditioned, and often purely voluntary type of response." The younger the infant, the less secondary behavior follows the startle response. As the infant matures, emotions begin to emerge.

Emotions and the Nervous System

As far as we know, individuals in all societies have the same emotions, although cultural differences may influence their intensity and expression. Charles Darwin believed facial expressions of the various emotions were universal and identical in all cultures.

It may be true: individuals in modern Western and Oriental cultures and primitive tribesmen in New Guinea show common facial expressions for basic feelings. Here facial expression is more reliable than words. Some languages have no single term for "anxiety" or for "depression." In translating Yoruba into English, for example, recourse has to be made to metaphors such as "the heart is weak" for depression, and "the heart is not at rest" for anxiety. One cannot be sure that these phrases correspond at all precisely with the English words.

Not only are some languages richer than others in the language of emotion; even within cultures members of some groups may have less well-developed abilities to recognize or to express nuances of feeling.

When it comes to fear, the "language of the body" is the most eloquent and universal. In *The Anatomy of Melancholy* (1621), Robert Burton described this well. Fear causes "many lamentable effects" in men, "as to be pale, tremble, sweat; it makes sudden cold and heat to come over all the body, palpitations of the heart, fainting. . . ." Burton knew fear but nothing about the nervous system. Much has been learned since his time, especially in the twentieth century.

Fear starts to run its course in the outermost layers of the brain called the cortex. When you hear the word "Fire!" in a theatre, it is registered in the sensory cortex. But then the "old brain" takes over—what biologist Paul McLean called the reptilian brain, which is submerged deep beneath the evolutionarily new cortex. The old brain does not "study the situation"; it acts, and acts the way it has for millions of years in thousands of species going back at least to the earliest vertebrates. It prepares the body to defend itself, to fight or flee.

Whichever happens, the messages from the old brain to the body are identical. They travel simultaneously over two routes: nerve fibers and the bloodstream (via hormones).

There are really two nervous systems. The *skeletal* nervous system activates muscles attached to bones (the skeleton) and makes it possible to run or fight—or just shake, as fear affects some people. The *autonomic* nervous system gives orders, via nerves, to internal organs like the heart and gut and instructs the adrenal glands—the little mushroom-like structures perched on top of the kidneys—to release adrenaline (epinephrine), a hormone that constricts blood vessels and stimulates the heart, among other actions. Autonomic means automatic. Once the cortex says "trouble!" the nervous system starts firing off muscles and glands. Although one can sometimes reason with skeletal muscles—appear calm, stop trembling—the old, primitive autonomic nervous system goes its own way.

The autonomic nervous system, in turn, has two parts, the *sympathetic* and the *parasympathetic* nervous systems, which have more or less opposite actions. The sympathetic nervous system is the emergency system; the parasympathetic system conserves energy for future emergencies.

Picture yourself in an emergency and you can easily predict whether your sympathetic or parasympathetic system is in command:

1. Your pupils dilate: the better to see with.
2. Your heart pumps faster: more blood means more energy.
3. You breathe faster: your body needs the oxygen.
4. You become pale: the blood is shunted from the skin to the muscles (the better to run with) and to the brain (the better to think with).
5. You stop digesting food: the blood is needed elsewhere, and if there is no tomorrow—disaster is striking—who needs to digest food?
6. You sweat: the sweat evaporates and cools off hot muscles.
7. Your hair stands on end. This does not help your survival chances at all, and merely shows how primitive and out-of-date our autonomic (automatic) responses sometimes are. When a cat's hair stands on end, the cat looks bigger and may scare away enemies, but not even the hair-

iest of us have that ability. "It is certainly a remarkable fact," Darwin wrote, "that the minute muscles, by which the hairs thinly scattered over man's almost naked body are erected, should have been preserved to the present day; and that they should still contract under the same emotions, namely, terror and rage, which caused the hairs to stand on end in those lower members of the Order to which man belongs."

All these things happen because the sympathetic nervous system takes over in emergencies. When the emergency has passed, the parasympathetic system returns to power: your pupils get smaller (the better to read with to learn about future emergencies) and your supper resumes its journey down the alimentary canal. Also, you start noticing the opposite sex again, sex being one of the first casualties of fear.

Sometimes the fear response outlasts the danger because of the residual effects of adrenaline. Impulses from the sympathetic nerve fibers release this amphetamine-like hormone from our adrenal glands and, circulating through the body, it does pretty much what the sympathetic fibers do: keep us primed for danger. The nerve fibers can be switched off immediately; it takes longer for adrenaline to clear from the blood.

This is how the body handles emergencies, but what about the *feeling* of fear? The feeling, of course, occurs in the brain. Not only that, it occurs in specific areas in the brain. Stimulate an electrode placed into certain brain areas (e.g., parts of the temporal lobe) and fear results. Electrodes inserted in other centers can produce feelings of anger, depression, and pleasure. One can also inject certain chemicals into these brain centers and produce specific feelings. When epileptics have seizures, they sometimes experience intense fear (or pleasure, or depression), presumably because the seizure originates in a particular feeling center. To imply, however, that the brain operates by push-pull buttons is misleading. Electrical and chemical stimulation of discrete areas in the brain sometimes produces specific emotions and sometimes does not. Brain function is highly plastic and, while there is some degree of localization, connections also tend to be diffuse and hard to trace.

SOME CELLS in the brain regions where fear can be produced with electrical impulses have receptors with a specific affinity for fear-reducing drugs such as Librium and Valium. The receptors are like locks and the drugs like keys that fit *only* these locks.

This was learned just recently and raises an interesting question. Before Librium was synthesized by a Polish chemist in 1955, there were no known chemicals in nature that fit these particular locks. What were they doing there? Was it possible that the brain itself produced chemicals like Librium which would be released in emergencies and for which the receptors were designed? Do we have built-in Librium? Do some of us have *more* built-in Librium than others, explaining why some of us are braver than others? Chapter 8 discusses this exciting new development in some detail.

THE CONNECTION between receptors, feelings, and physiology is intricate indeed. Controversy goes back over a century about which comes first: the feeling of fear or the physiological changes. William James, the Harvard philosopher-psychologist, believed the changes came first. The sense organs and brain perceive a danger. Palpitations, rapid breathing, a tensing of muscles follow. Only then does the person experience fear, because he recognizes the physical changes signifying danger.

This sounds academic—who cares which comes first?—except for one consideration: if the physiological changes are primary and if they could be eliminated in situations where they are not needed, this might reduce the sensation of fear.

To a certain extent this can be accomplished, affording some support for the James-Lange theory (Lange was an associate of James). Drugs that block palpitations have been given to anxious patients and have made them feel less anxious. Lactic acid, a chemical that produces palpitations, also produces anxiety, especially in anxiety-prone people.

Further support for the James-Lange theory comes from the finding that people paralyzed from the waist down experience fear and anger, whereas those paralyzed from the neck down experience these emotions to a much lesser degree. Quadraplegics describe themselves as acting emotionally but not feeling emotional. As one put it, "I *say* I am afraid, like when I am going to a stiff examination at school, but I don't really *feel* afraid, not tense and shaky with that hollow feeling in my stomach, like I used to."

If the James-Lange theory is true, then one would expect that individuals with high levels of anxiety or those susceptible to anxiety disorders, such as phobia, would have unusually reactive autonomic nervous systems. There is some evidence this is true. In 1937, investigators reported that patients with "anxiety states" had faster heart rates than did normal subjects even when not anxious. This was confirmed by some studies but not others. One study found that anxious patients had normal heart rates but were more *aware* of their heart rates than normal subjects. The investigator concluded that perhaps the anxious person is simply more sensitive to normal physiological responses. Once the physiologic responses become accentuated, however, individuals susceptible to anxiety require a longer time than normal persons to return to normal.

In short, in order to prove the James-Lange theory, investigators would have to show that rapid heart rate, increased breathing, and other physiological changes routinely preceded the subjective state of anxiety as opposed to the possibility that physiologic fear responses are really consequences of high levels of anxiety. The evidence remains contradictory.

ELIMINATING FEAR—through modern chemistry or otherwise—would of course be disastrous. Like such drives as hunger, thirst, and sex, emotions have survival value for both the individual and the species. Psychologists have devised a U-shaped model demonstrating a relationship between fear and performance. In the absence of fear, performance is relatively poor; with some degree of fear, performance improves; with excessive fear, performance once again is poor.

Eliminating anxiety is another matter. As defined in Chapter 1, anxiety has no more survival value than a tension headache. Its elimination would be a blessing.

The emotional reaction to a threatening situation depends on the *interpretation* an individual makes of the situation. If the danger is interpreted as one that can be mastered by attack, the emotion will be anger and certain physical changes follow from that interpretation. In anger, the lids of the eyes are narrowed to restrict the vision to that part of the environment the organism seeks to attack. If the situation is seen as one that cannot be overcome by assault but can be avoided by flight, the emotion will be fear: the eyelids open wide to give the organism opportunity to see every possible route of escape.

These oversimplifications, which admit many degrees of shading, bring us to the story of Tom—one of the classic case histories in modern medicine. As a boy, Tom (his last name is never provided) drank some boiling hot chowder, which caused his esophagus to close. To permit him to eat, a surgeon made a passageway (a fistula) from his skin to his stomach. For many years afterwards he fed himself by means of a funnel through the fistula.

Looking through the funnel, one could observe the state of Tom's stomach. Two physicians named Stewart G. Wolf and Harold G. Wolff took advantage of the opportunity and spent a good deal of time looking into Tom's stomach. They discovered that when Tom feared some objective danger, his gastric activity *decreased:*

> Sudden fright occurred one morning . . . when an irate doctor, a member of the staff, suddenly entered the room, began hastily opening drawers, looking on shelves, and swearing to himself. He was looking for protocols to which he attached great importance. Tom, who tidies up the laboratory, had mislaid them the previous afternoon and he was fearful of detection and of losing his precious job. He remained silent and motionless and his face became pallid. The mucous membrane of his stomach also blanched and remained so for five minutes until the doctor had located the objects of his search and left the room. Then the gastric mucosa gradually resumed its former color.

By contrast, Tom showed *increased* gastric activity when he became anxious:

> The most marked alterations in gastric functions . . . were associated with anxiety provoked by our failure to inform the subject how long he might expect an income from the laboratory. He had been receiving government aid prior to his employment with us, and the rise in his family's standard of living since his new job meant a great deal to him. The subject of how long his job would last had come up in a discussion between his wife and himself the previous evening. He decided to inquire about it the next morning. Both he and his wife were so anxious about the answer, however, that neither of them slept at all. The next morning the values for vascularity and acidity were the highest encountered in any studies. . . . [Increased vascularity means reddening of the stomach lining.]

This was the first evidence that fear and anxiety produce different physical responses. Wolf and Wolff's interpretation of this case has been debated, but it may be correct with regard to mild anxiety. Severe anxiety, however, produces physical reactions identical to those experienced by persons confronted with actual danger, although the danger, in anxiety states, is formless and obscure.

4

Innate Fears

Alas, our frailty is the cause, not we! For such as we are made of, such we be.

Shakespeare
Twelfth Night, II, ii, 32

Some fears are innate. There is no other explanation for an adult chimpanzee, seeing a snake for the first time twenty yards away in a zoo, going into a panic. People also have innate fears: fear of strangers, fear of being stared at, fear of falling into space. These fears even follow a kind of biological timetable. At six months, the infant is frightened by loud noises and sudden movements. At age three he is frightened by strangers; at five by animals. Fears of open spaces and social situations occur (*if* they occur) much later: in adolescence or early adulthood. There is a remarkable consistency across cultures in the timing of these "normal" fears, strongly suggesting biological factors.

Innate does not necessarily mean inherited. Before a person is born, he spends nine months in the womb, perhaps the most important nine months in in his life. Experiences in the womb powerfully influence postnatal appearance *and* behavior. For example, if a female monkey fetus receives male sex hormones at a crucial period of development, the female newborn not only has masculine features but, later, behaves sexually as a male.

In the seventeenth century, Descartes observed that if a pregnant woman smells roses or is frightened by a cat, her unborn child will also smell roses and may have an aversion to roses or cats "imprinted in his brain to his life's end." Far-fetched as this now seems, no one doubts that what passes through the bloodstream of the mother also passes through the bloodstream of the newborn, and that fear involves molecules in the blood. Intrauterine experiences may have no relevance to postnatal fears, but, again, they might, and Descartes' point that "innate" is not always "genetic" is well to remember.

IN ANY CASE, there is no question that animals inherit fears. Consider the famous hawk experiment:

> If you pass a V-shaped shadow sideways or backwards over a baby chick in a lab, nothing happens. But if you move the same shadow forward, the baby chick goes into a panic. He has inherited the genetic message—*hawk, danger*—for a certain-shaped shadow, even though he is only a couple of days old and has not talked to his mother and may never see a hawk. According to Konrad Lorenz, the father of ethology, the V-shaped shadow is a releasing mechanism for an innate fear response.

Identifying the precise "releasing mechanisms" for innate fear responses in humans is often impossible, but the evidence that humans *have* innate fears is impressive. In the following eight types of fears, innate factors appear important.

1. Fear of Darkness

Almost all children are afraid of the dark after the age of three. Apparently what the child really fears are the horrendous creatures—serpents, monsters—conjured up by the imagination. Children in different cultures imagine the same monsters; it is hard to attribute this to upbringing.

2. Fear of Strangers

This occurs normally in infants 6 to 12 months old. Fear of strangers and smiling are related. By two months the baby smiles indiscriminately. After six months, the baby smiles only at familiar faces, particularly familiar combinations of forehead, eyes, and nose (incidentally, sex or race do not matter). Seeing both eyes appears to be necessary for the smile response. Strangers elicit screams rather than smiles, particularly if they have large eyes (perhaps wearing spectacles) and large teeth.

Smiling and fear of strangers are universal in 6- to 12-month-old babies. Smiling certainly makes sense. It elicits parental care. It is the first evidence of social interaction in humans, and has survival value. Other mammals and birds also fear strangers. The chimpanzee, for example, begins fearing strangers at about the same time in life as the human infant.

Innate fears are modified by early experience. If the infant has been exposed to many people, he will not be as shy as he would be if he had grown up with only a few (just the family, say). In the latter case, a stranger may produce a response as violent as pain does.

The response is at full strength on first exposure, showing that it is not learned. However, an experiential factor still exists. If the chimpanzee has been reared in darkness, or if the human child has had congenital cataracts until after 6 or 8 months and has then been operated on, the fear response does *not* occur on first exposure. The infant must first become visually familiar with a small group of persons; only afterwards does a stranger evoke fear.

From these observations the psychologist Donald Hebb concluded that fear of the strange is elicited by events that combine familiar and unfamiliar elements, producing uncertainty and confusion. He gives the following example:

> Dr. R. and Mr. T. are regular attendants in a chimpanzee nursery; the infants are attached to both and evenly welcome being picked up by either. Now, in full sight of the infants, Dr. R. puts on Mr. T.'s coat. At once he evokes fear reactions identical with those made to a stranger, and just as strong. *An unfamiliar combination of familiar things, by itself, can therefore produce a violent emotional reaction.* [Hebb doesn't explain why this happens, but it undoubtedly involves an emergency response to the unexpected.]

In situations where previous learning is an element in the fear response, years may pass before it can occur. Fear of the dark, for example, does not occur until age three or later, allowing time for the brain to be capable of imagining monsters.

3. Fear of Dead or Mutilated Bodies

This fear can be reduced by repeated exposure—consider the soldier, the medical student, the undertaker's assistant. But as an initial visit to the dissecting room shows, the reaction is often powerful on the first exposure, indicating an innate factor. It appears also in milder form with exposure to persons with mutilating injuries (especially to the face).

Again, the reason may be that dead and mutilated people look almost real, but not quite.

Chimpanzees feel comfortable when they are with other chimpanzees but when shown the sculptured head or death mask of a chimpanzee they become panic-stricken The response is age dependent. Two-year-old chimps ignore the model head, looking only at the experimenter; five-year-olds are fascinated, coming close to stare persistently and excitedly at it; older animals, nine and over, show outright terror, screaming with hair on end. Even with a wire cage between them and the clay model of a chimpanzee head, none of the older animals will approach it.

4. Fear of Snakes

Adult chimpanzees fear snakes. They fear snakes even if they have never seen a snake before. The fear is inherited.

Do humans have an inherited fear of snakes? Maybe. If so, it comes with age: small children do not usually fear snakes. It may not develop at all if the child has frequent opportunities to play with snakes; these children learn *not* to fear snakes. But if you show a snake to city-bred adolescents who have never seen a snake and tell them it is harmless, they may say they believe you, but still keep their distance.

Perhaps the city-bred kids have *learned* that snakes are dangerous from reading or from their parents. Still, adolescents learn about a lot of dangerous situations—such as taking drugs and driving too fast—and this does not seem to discourage other forms of risk-taking.

Also, it has been suggested that perhaps chimpanzees and city children exposed to a snake for the first time are not frightened by the snake itself (they do not have a small portrait of a snake labeled "danger" in their brain) but rather are fearful of the writhing movements of the snake. There is, in fact, evidence that chimps fear strange *moving* objects more than similar stationary objects.

Still, fears of strangers, serpents, and bats are pervasive and most authorities suspect a hereditary factor. One early authority, Stanley Hall, in 1897, noted that

> serpents are no longer among our most fatal foes. Most of the animal fears do not fit the present conditions of civilized life; strangers are not usually

dangerous nor are big eyes and teeth . . . yet, the intensity of many fears, especially in youth, is out of all proportion to the exciting cause. First experiences with water, the moderate noise of the wind, distant thumder, etc., might excite faint fear, but why does it sometimes make children frantic with panic?

Adam Smith says it is in the genes:

> There was a time when humanity slept by the fire and predators really were there in the dark. One cell carries the instructions on how to make a whole new person: let's see, make the eyes brown, make the ears thus, make the nose thus, let's set the trigger for adolescent growth at 12.2 years, and oh, yes, let's throw in the message about the bears, predators by the fire, for ages five to nine.

5. Fear of Dark Woods

When people started using LSD in the early 1960s, they shared their experiences and learned they were often the *same* experiences. One common LSD-produced fear was of entering dark woods and not finding the way back. As one user said, "Everybody had a dark woods in the middle of their head, and sooner or later they got around to it." So when Bob Dylan sang about the foggy ruins of time, past the frozen leaves, the haunted, frightened trees, he touched a universal of sorts.

The feeling goes back at least to the time when dark woods really were dangerous and that is where most people lived, a time resurrected by fairy tales and Walt Disney movies. Whether the fear is innate or learned (or both), dark woods are a pervasive symbol for lurking monsters and the unknown. Dante opens *The Divine Comedy* with "In the middle of the journey of our life, I found myself in a dark wood," and a terrifying place it turns out to be. Fear of dark woods goes beyond simple fear of the dark. Adults otherwise fearless in the dark quake at entering a dark wood. Innate fear can be highly specific.

6. Fear of Heights

Even goats fear heights. This was shown in a classic experiment. A board was laid across the center of a sheet of glass. On one side of the board a sheet of patterned material was placed flush against the undersurface of the glass, giving the glass the appearance of solidity. On the other side a sheet of the same patterned material was laid on the floor one foot below the glass. Placed on the "deep" side of the glass, newborn goats froze; they relaxed again only when removed to the "shallow" side. Human infants behaved the same way, crawling readily on the "shallow" side of the board and avoiding the "deep" side. Fear of a receding edge remains among adults on the edge of a precipice, with the feeling of being drawn down and a protective reflex to withdraw from the edge.

7. Fear of Being Looked at

Eyes are probably the first thing babies notice. They are small, colorful, move, and reflect light. Two eyes, as noted, are required to elicit the first social response, smiling. The first drawings of children often are of big round heads with eye spots. Legs and arms are added later. "Fear of two staring eyes is ubiquitous throughout the animal kingdom, including man," according to Isaac Marks, a leading authority on phobias. Belief in the power of the look seems universal and independent of culture. In man, large staring eyes are used in defensive magic. Many species of mammals and birds use eyes and eye markings as threat displays and defense against attack.

Stare at a monkey in a cage and you produce erratic changes in the monkey's brain waves as well as its behavior. Stares produce discomfort in most of us, and the cause seems to be inheritance as well as early life experiences.

8. Fear of Novelty

People and animals have mixed responses to novelty. The novel can produce fear but may also be sought out. Novelty can attract and repel in turn, as demonstrated by Lorenz's wonderful description of a mixed-up raven.

> A young raven, confronted with a new object, which may be a camera, an old bottle, a stuffed polecat, or anything else, first reacts with escape responses. He will fly up to an elevated perch and from this point of vantage, stare at the object . . . maintaining all the while a maximum of caution and the expressive attitude of intense fear. He will cover the last distance from the object hopping sideways with half-raised wings, in the utmost

INNATE FEARS

readiness to flee. At last, he will deliver a single fearful blow with his powerful beak at the object and forthwith fly back to his safe perch. If nothing happens he will repeat the same procedure in much quicker sequence and with more confidence. If the object is an animal that flees, the raven loses all fear in the fraction of a second and will start in pursuit instantly. If it is an animal that charges, he will either try to get behind it or, if the charge is sufficiently unimpressive, lose interest in a very short time. With an inanimate object, the raven will proceed to apply a number of further instinctive movements. He will grab it with one foot, peck at it, try to tear off pieces, insert his bill into any existing cleft. . . . Finally, if the object is not too big the raven will carry it away, push it into a convenient hole and cover it with some inconspicuous material.

Monkeys behave in a similar manner. Confronted with a strange object, a monkey freezes and stares at the object from a distance. After a while the animal tentatively approaches the object, looking at it, sniffing, touching, and finally handling it. After an hour or so, the animal returns to staring fixedly at the object.

Do our children do the same? In their own way, yes.

Individual Variation

Individuals and species vary enormously in their response to fear-provoking situations, even when the response appears largely innate. This variability can be traced to a combination of four factors:

The Evolutionary Scale

Psychologist Donald Hebb wrote:

> As we go from rat to chimpanzee (from lower to higher animals), we find an increasing variety in the causes of fear. Pain, sudden loud noise and sudden loss of support are likely to cause fear in any mammal. For the laboratory rat we need add only strange surroundings, in order to have a list of things that disturb the animal under ordinary circumstances. With the dog, the list becomes longer: we must add strange persons, certain strange objects (a balloon being blown up, for example, or a large statue of an animal) or strange events (the owner in different clothing, a hat being moved across the floor by a thread which the dog does not see). Not every dog is equally affected, of course, but dogs as a species are affected by a much wider variety of things than rats.

Monkeys and apes are affected by a still wider variety, and the degree of disturbance is greater. Causes of fear in the captive chimpanzee make up an almost endless list: a carrot of an unusual shape, a biscuit with a worm in it, a rope of a particular size and color, a doll or a toy animal . . .

And humans? Evidence from ethology suggests that the more intelligent the animal, the greater the susceptibility to baseless fears. If Part III of this book is an indication, humans must surely surpass any other species in range and variety of baseless fears.

Genetic Differences

Species differ greatly in fearlessness; lions and tigers are more fearless than deers and rabbits. Moreover, animals can be inbred to show greater or lesser fear responses to specific stimuli and to be more or less fearless in general, as shown by inbreeding of dogs. Among humans, identical twin children sharing the same genes fear strangers about equally; fraternal twins, with different genes, have differing degrees of fear.

Imprinting

Experiences in infancy modify innate fear responses. This is called imprinting and it can occur very early in life. For instance, chicks hatched from eggs incubated in darkness differ from chicks hatched from eggs incubated in light: they are less fearful! From the moment of conception, genetic and environmental influences interact. The question "What is inherited? What is learned?" is often unanswerable.

Stimulus Characteristics

In animal research, fear is inferred from behavior. If an animal approaches an object, it is "unafraid"; if it avoids an object it is "afraid." Approach-avoidance studies have produced the following generalizations:

(a) Animals fear (avoid) loud, irregular noises and large objects moving toward them at high speed (particularly objects with sharp, angular corners).
(b) Animals are unafraid of (approach) small, rounded objects that move slowly and make soft, regular noises.

In human terms, the boss who produces fear in his employees is the burly fellow with craggy features and a loud staccato voice who comes at you full-steam, lays both hands on your shoulders, and says . . . it hardly matters what he says. It could be "Nice day!" and the pupils of the employee will still dilate, his heart pound, and the adrenaline flow.

Summation

Some situations are feared more than others, not because of any single factor described above but from a combination of factors which, in a given individual in a particular situation, produces fear. Genetically determined fear may be "latent," expressed only with the addition of imprinting or later stressful life experiences or in the presence of unusually intense stimuli. This is called summation. The fearful person is almost never aware of summation: the complex, intertwined reasons why he is afraid.

5

The Role of Conditioning

> Observation discloses in the animal organism numerous phenomena existing side by side and interconnected now profoundly, now indirectly, or accidentally. Confronted with a multitude of different assumptions the mind must *guess* the real nature of this connection. Experiment, as it were, takes the phenomena in hand, sets in motion now one of them, now another, and thus, by means of artificial, simplified combinations, discovers the actual connection between the phenomena.
>
> I. P. Pavlov
> *Experimental Psychology and Other Essays*, Pt. X, Essay 3
> (tr. by S. Belsky)

From his observations of phenomena "existing side by side," I. P. Pavlov discovered the conditioned reflex. The great Russian physiologist, who died in 1936 after nearly seventy years of productive work, found that just as a knee jerk or eye blink could be caused by "conditioning," so could fear. Some knowledge of conditioning is essential to understanding fear and anxiety.

What is conditioning? The term refers to the production of physiological or psychological responses by the repeated presentation of a stimulus that ordinarily would not produce the response but eventually does so by the repetitive pairing of the stimulus with a response that occurs naturally. Conditioning is explained most easily by illustration. The classic illustration, of course, involves Pavlov's dog.

Pavlov repeatedly rang a bell just before feeding the dog. After a time the dig salivated whenever the bell rang, whether it received food or not. The dog was *conditioned* to salivate.

Four terms are needed to understand conditioning:

- *Unconditioned stimulus*—e.g., hunger
- *Unconditioned response*—e.g., salivation
- *Conditioned stimulus*—e.g., the bell
- *Conditioned response*—e.g., salivation

Before conditioning can occur, there must be an organism and the organism must have "drives," such as hunger. Drives evolved in the interest of survival of the individual and the species. Drives have one aim: their own abolition. Saliva digests food, temporarily abolishing hunger.

By repeatedly pairing a neutral stimulus (bell) with a drive (hunger), a conditioned response (salivation) occurs and this becomes independent of the original drive. The bell rings, the dog salivates, whether hungry or not. The dog has "learned" something in much the same way we learn to drive a car, namely, by acquiring a set of conditioned responses. (Learning and conditioning are used synonymously in this chapter, although learning embraces other processes than conditioning.)

We constantly form new conditioned responses and discard old ones, and usually are never the wiser.

Habits are conditioned responses. Tying our shoelaces, brushing our teeth, waving at a friend: all are habits, all learned, all more or less automatic, all rooted in a drive with survival value (staying warm, being attractive, maintaining allies).

Conditioned stimuli become paired (and unpaired) to drives endlessly; and although there is a limited number of drives (hunger, thirst, sex, among others), literally *anything* can become a conditioned stimulus.

Say it thunders every time you eat marshmallows. After a few days, thunder makes you hungry for marshmallows. You may even *crave* marshmallows. Your marshmallow addiction results from a conditioned response (hunger) to a conditioned stimulus (thunder) that became paired through happenstance. Thus, when one expert says that addictions are biological and another says they are learned, both are right (and neither usually mentions happenstance).

In the same sense, few fears are purely innate or purely learned—at least in higher animals. Four centuries ago, Sir Francis Bacon wrote: "Men fear Death, as children fear the dark; and as that natural fear in children is increased with tales, so is the other." Tales become conditioned stimuli, but a biological factor—the innate fear, the unconditioned response—must be present for conditioning to occur.

Words become conditioned stimuli. Think of "I love you" or "You're fired" or "I'll kill you" and the point is obvious.

Nonverbal symbols become conditioned stimuli. A swastika once inspired alarm in entire communities. Secondary sexual characteristics in the human female—breasts, bottoms—invoke unconditioned responses in the human male, but the size of breasts and bottoms that invoke these responses changes from era to era, showing how condition-bound unconditioned responses become.

When the conditioned response is especially strong, a single pairing with a neutral stimulus may produce a lasting conditioned response. A survivor of a plane crash may panic at the mention of the word "plane." A mouse may once become ill from poisoned food in a cupboard and never go back to that cupboard again. This is called one-trial learning.

Conditioned stimuli tend to generalize: things remind people of things. Mark Keller, writing about alcoholism (where conditioning plays an important role), describes the phenomenon aptly:

> For the alcoholic, there may be several or a whole battery of critical cues or signals. By the rule of generalization, any critical cue can spread like the tentacles of a vine over a whole range of analogs, and this may account for the growing frequency of bouts, or for the development of a pattern of continuous inebriation. An exaggerated example is the man who goes out and gets drunk every time his mother-in-law gives him a certain wall-eyed look. After a while he has to get drunk whenever any woman gives him that look.

A little boy is frightened by a mouse, later fears any small furry animal, and as a teenager develops a phobia toward all animals. The "rule of generalization" is important in the development of phobias.

CONFUSING STIMULI can also produce "neuroses" like phobias. Pavlov created the first neurotic dog (at least in a laboratory). He did it as follows.

A circle was projected on the wall in front of the dog and then the dog was given food. When an oval was projected on the wall the dog was not given food. At first the circle and oval were quite different, the ratio of length to breadth in the oval being 2:1. The dog secreted saliva to the circle, not to the oval. Then the ratio was decreased. When it became 9:8 (the oval was now nearly a circle), the dog had a nervous breakdown. It howled, defecated, and refused to enter the experimental room.

Even when the experimenter restored the symbols to their original shapes—perfect circles, unmistakable ovals—the dog remained wacky. "His neurosis seems permanent!" marveled Pavlov.

Remember: the dog had not been punished; he was never in pain. He simply was a victim of conflicting signals. It happens to people all the time.

SO FAR, WE have been talking about Pavlovian or "classical" conditioning: the pairing of a neutral stimulus with an unconditioned response. The response—salivation, rapid heartbeat, fear, sexual arousal—is largely involuntary.

There is another type of conditioning called "operant" conditioning where the animal or person learns to "operate" in ways calcuated to produce pleasure or avoid pain. The behavior is often repetitive and predictable—this is why it is called "conditioning"—but the person has more control over his behavior than he does of his salivary glands, and thus operant conditioning is more voluntary and thought-out than classical conditioning. The person is more his own operator, so to speak.

Classical conditioning and operant conditioning are closely related. The first often procedes the second. For example:

Put a rat on an electrified grid. Flash a red light and, one-half second later, give the rat an electric shock. After doing this for a time, the red light alone will produce freezing, jumping, or other conditioned fear responses. Classical conditioning has occurred.

Then give the rat an opportunity to avoid or escape the shock. Eventually the fear response to red lights will disappear. If, for example, the rat learns that by pressing a bar when the light flashes, the shock will not occur, the rat after a while will become a skilled bar-presser. He has learned to avoid punishment by his behavior. Should he become lazy or forget to press the bar, another jolt serves as a reminder. The jolt is called a "negative reinforcer."

How many negative reinforcers in the form of memoranda flow daily from bosses to employees?

It also works the other way. If an animal learns to press a bar to receive food, the food, a reward, serves as a positive reinforcer, and the rat becomes a happy and hardworking bar-presser. A nice word from a supervisor often has the same effect: the recipient is not only grateful but works harder.

Behavior is powerfully shaped by negative and positive reinforcers. We simply learn which actions bring rewards and which bring punishments and behave accordingly. However, it is not always that simple.

SOMETIMES THE *same* action produces both rewards and punishments. It happens all the time in real life and is easy to show in the laboratory.

A rat is first taught to press a bar to avoid shock and later taught to press the same bar to obtain food. You now have a very confused rat: another case of conflicting signals, now compounded by conflicting reinforcers. Clearly a set-up for "emotional problems"—in rats and people. You love your wife but she is a bad housekeeper: stay or leave? (or both—take up golf).

Conflict is the rule of life. This is partly the basis of the popularity of sedatives and alcohol. Both reduce conflict—temporarily. Again, you can show it in the lab.

Cat food is placed at the end of a tunnel. A green light flashes when the food is available. At the end of the tunnel is also an electrified grid. A red light flashes when the grid is activated. Both lights become conditioned stimuli: the cat salivates at green and is terrified by red.

Place a very hungry cat in a tunnel and flash both lights simultaneously. Something called an approach-avoidance conflict results. The cat moves a certain distance down the tunnel toward the food but freezes at the expectation of shock. Hunger versus fear. How far the cat goes down the tunnel will depend on which is stronger: hunger or fear.

At this point give the cat alcohol or a barbiturate. The fear is reduced but not the hunger. He now approaches more than avoids. Reducing the fear pharmacologically has temporarily resolved the conflict.

It happens all the time in bars. The man on the bar stool, the woman sitting at the table . . . the third drink, the fourth . . . the approach. "Love casts out fear," as Aldous Huxley said. Lust and alcohol combined *really* cast out fear.

Animals quickly learn the fear-reducing effects of alcohol and barbiturates. In approach-avoidance situations, they will self-administer both drugs. People learn, also. There is approximately one bar for every grocery store in most towns.

Alcohol reduces fear but the fear does not stay reduced after the alcohol is gone. There are other ways to reduce fear where the fear does stay reduced. They are discussed briefly now and will be explored much more fully in the chapter on the treatment of phobia.

Deconditioning (Overcoming Fear)

There is an old saying in psychology that *anything learned can be unlearned.* In this message lies hope for fearful humanity, if only more was known about how to unlearn.

It is known that conditioned stimuli lose their strength (extinguish) if not maintained by at least an occasional reward or punishment.

Pavlov's dog eventually stops salivating on hearing a bell if presentation of food *never* follows the bell. Fear of snakes (even if partly innate) is overcome with frequent exposure to snakes. A flashing neon bar sign remains a conditioned stimulus signaling "thirst" only as long as it regularly produces the conditioned response of entering the bar and ordering a drink. If the response is suppressed for a sufficient period of time, the stimulus loses its grip.

Extinguishing thirst-producing stimuli, by the way, is an essential part of the treatment of alcoholism, and the difficulty in treating alcoholism is that "sufficient" time may be a long time indeed.

Suppression of the conditioned response must be unrelenting to neutralize the conditioned stimulus. Erratic responding, if anything, strengthens the "habit." Slot-machines—all forms of gambling, for that matter—are popular and even become addictive because the payoff is erratic. "Maybe this time I'll win" keeps the gambler gambling. It is a thought that never occurs in situations when you never win. Nor does it occur in situations where you *always* win. In the latter, people sometimes stop what they are doing out of sheer boredom.

Conditioned fear stimuli resist extinction under two circumstances. First, if the person avoids exposure to the feared situation, he never has an opportunity to unlearn the fear. Second, if alcohol or drugs are used to reduce the fear, they do so, as mentioned, only temporarily. As all drinkers know, the fear returns—often redoubled—on the morning after.

Nonresponse to fearful stimuli and familiarization with fearful situations are ways people handle ordinary fears all the time. Familiarization often is vicarious. Hearing someone describe a fear they have had and how it was overcome can be reassuring. Familiarization with fear through shared experiences also takes the form of gossip, which can be a healthy habit rather than a minor vice. Just knowing that others have fears helps. Some people—a stoical woman, a macho man—may *seem* fearless, making others feel weak and cowardly, but it is often camouflage. Movies, plays, and novels all depict fear in others and help us handle our own.

THERE IS ANOTHER principle in psychology that helps people deal with fear. This is the "law of reciprocal inhibition."

Reciprocal inhibition simply means that you cannot experience two opposing emotions simultaneously. You cannot be afraid and feel sexy at the same time. (Think so? Try it.) You cannot feel affectionate toward your lover and be mad at him or her simultaneously. Angry men forget their fear in combat. Even Pavlov's neurotic dog showed less fear when sexually stimulated.

Knowingly, unknowingly, people use this principle to combat fear all the time. The fear of flying is overcome by dwelling on the good time awaiting at the other end. Claustrophobes, heart pounding, take the elevator up to the office thinking fervently of the things they will buy with their pay. Fear and socializing do not go together, perhaps explaining why some people are so gregarious.

People learn what comforts them, and learn that what comforts them reduces fear. A little girl puts her thumb in her mouth every time she's frightened. As long as she has her thumb in her mouth she seems impervious to fear. Another child fears certain objects while sitting on the floor but not when sitting in his familiar high chair.

The thumb or high chair for these children is a *soteria*, a term for any object or situation from which people derive disproportionate comfort. A soteria is the opposite of a phobia. Examples are stuffed animals and toys which children carry around with them, and charms which many adults wear. For the youngster on the streets, the loud stereo he is lugging around may in fact be as much a soteria as it is a radio.

Some people afraid of fainting carry a bottle of smelling salts, although perhaps they have never fainted in their life. Others have aspirin or Valium with them at all times and never take them.

Eating reduces fear. Chimpanzees lose their fear of snakes if bananas are placed on the lid of a glass box containing a snake. Joseph Wolpe, who devised a successful treatment for phobias, discovered the treatment from a study involving food and cats. He put the cats in a cage containing food. Whenever they walked toward the food, they were shocked. Pretty soon the cats decided they would rather go hungry than be shocked, and developed a phobia, so to speak, about food in cages. The phobia persisted even when the shocks were discontinued.

Wolpe found a way to "cure" the phobia that he later applied to human phobias. He first placed the food some distance away from the cage, sufficiently far for the approach tendency due to hunger to overcome the avoidance tendency due to fear. Gradually, the food was placed nearer the cage and finally inside of it, until eventually the cats entered the cage and ate the food. More will be said about Wolpe and his phobia cure in Chapter 18.

6

Psychological Theories

> There is no question that the problem of anxiety is a nodal point at which the most various and important questions converge, a riddle whose solution would be bound to throw a flood of light on our whole mental existence.
>
> Sigmund Freud
> *Introductory Lectures on Psychoanalysis*

> What does Freud know?
>
> J. B. Watson
> Letter to a friend

There are two types of psychologists—those who talk about the mind and those who don't. Sigmond Freud and Rollo May are examples of the first; John B. Watson exemplifies the second.

Freud was a neurologist who specialized in treating middle-class neurotics, but his fame came from his writings in psychology. He was a cartographer of the mind par excellence. His map had three parts: the id, the ego, and the superego. The id was the source of primitive instincts; the superego was the stern voice of conscience; and the ego was the arbitrator who tried to reconcile the demands of id and superego and get along in the world. Anxiety was a "signal" that a state of war existed between the three constituencies.

John B. Watson, the father of behaviorism, reduced behavior to stimuli and responses and said that whatever went on inside the head was unknowable (a "black box"). He rejected Freud's theories *in toto*.

Whichever you accept—the theories of Freud and his followers, or the observations of the behaviorists—the two have dominated twentieth century psychology. Anxiety was central to Freud's theories. Watson, rejecting mind, also rejected anxiety as a theoretical concept. However, the behaviorists' studies are as relevant to an understanding of anxiety as the theories of Freud.

Freud's theories about anxiety passed through two stages. First he believed that anxiety came from sexual frustration—what he called undischarged libido. He blamed not only sexual abstinence but also masturbation and coitus interruptus, all of which he considered unsatisfying substitutes for coitus uninterruptus.

Later he saw anxiety as an anticipation of danger—the danger being the "return of the repressed." As described in Chapter 1, Freud believed that small children lusted for the parent of the opposite sex and feared retaliation from the parent of the same sex. This created a conflict so distressing that it was actively thrust out of conscious awareness, repressed by the ego. The id and superego were locked in combat, with the ego sensing danger but unable to trace its source.

Other psychologists have redefined the nature of the conflict without abandoning the idea that anxiety is an anticipation of danger that cannot be identified. Among them is Rollo May. A psychologist and popularizer of existentialism, May accepted Freud's ideas but added to them some of his own drawn from existentialism. Except for Freud, perhaps, no other psychologist has given such extensive attention to anxiety.

May agreed with Freud that the difference between fear and anxiety is that fear is a reaction to a specific danger while anxiety is unspecific, vague, "objectless." In his book, *The Meaning of Anxiety*, he says that anxiety is not necessarily more intense than fear. Rather, it attacks at a deeper level, at the "core" or "essence" of the personality. He then proposes the following definition: "*Anxiety is the apprehension cued off by a threat to some value that the individual holds essential to his existence as a personality.* The threat may be to physical life (the threat of death), or to psychological existence (the loss of freedom). Or the threat may be to some other value one identifies with one's existence (patriotism, the love of another person, 'success,' etc.)."

The occasions for anxiety vary as widely as the values on which people depend. Since anxiety attacks the foundation of the personality, the individual cannot "stand outside" the threat, cannot objectify it. "Therefore, one is powerless to take steps to confront it. One cannot fight what one does not know."

The inevitability of death is an "occasion" for anxiety. "Since anxiety threatens the basis of selfhood, it [involves] the realization that one may cease to exist as a self. One is a being, a self; but there is at any moment the possibility of 'not being.' Death, fatigue, illness . . . are all illustrations of nonbeing. Death involves the dissolution of the self, not merely physical death. It involves loss of psychological or spiritual meaning identified with one's existence as a self—the threat of meaninglessness."

May uses a military analogy: "Battles on various segments of the front lines represent specific threats. So long as the battle can be fought out on the periphery, so long as the dangers can be warded off in the area of the outer fortifications, the vital areas are not threatened. But when the enemy breaks through into the capital of the country, when the inner lines of communication are broken and the battle is no longer localized—when, that is, the enemy attacks from all directions and the defending soldiers do not know which way to march or where to take a stand, we have the threat of being overwhelmed." This threat to basic values, the "inner citadel" of the personality, is what we experience as anxiety, according to May.

RETURNING TO the behaviorists, John B. Watson never formulated a specific theory of anxiety. Rather, he discovered ways to produce anxiety experimentally. There is no better way to compare Freud and Watson than to examine their ideas about phobia, one of the major anxiety disorders.

Freud and Watson on Phobias

Phobias do not occur at random. Any explanation for them must answer these questions:

1. Why do some people develop phobias while most do not?
2. Given so many things to fear, why does a person become phobic about a *particular* thing?
3. Why are some phobias especially common?
4. Why do certain phobias occur at certain ages or stages in life?
5. Why do women develop phobias more often than men?

There are many theories about the cause of phobias, but none satisfactorily deals with these questions. The cause of phobias remains unknown.

The theories are interesting, however, and some may even be true, in part. Theories are often viewed as mutually exclusive, but this is a mistake. Conceivably, if some genius could identify the true parts and synthesize them into a theory that could be tested scientifically, the cause or causes of phobia might become clear. No synthesizer has appeared, and we are left with theories bumping shoulders but rarely speaking to each other.

The theories of Freud and Watson are presented here in more or less their original form. As the years passed, both theories were modified somewhat by their originators and followers. However, the modifications did not modify much and will only be touched on slightly.

Both theories were inspired by little boys.

Little Hans was a five-year-old when Freud studied his case. Hans had a horse phobia. After seeing him once and talking with the father, Freud traced the phobia to this sequence of events:

1. Hans lusted for his mother.
2. He was jealous of his father, a rival with a larger penis.
3. He could not express his jealousy because of fear, specifically fear that he would be punished by the father for lusting for the mother. The punishment would take the form of castration (a punishment he assumed was common, based on his observation that little girls lacked penises, presumably because they, too, had lusted for a parent).
4. The fear was so great that he actively "forgot" *(repressed)* the whole business. For further protection, his *unconscious mind* performed the following maneuvers:
5. He *projected* his hostility onto the father: the father became the aggressor. Then he *displaced* his aggressive feelings from the father to something safer, more neutral: horses.

Horses became the objects to be feared and avoided. Horses were a *symbol* for the father, but Hans did not know this. He certainly did not connect his fear of horses with sex—incestuous sex, at that!

In the Little Hans analysis are all the ingredients of the psychoanalytical theory of phobias: sexual feelings toward the parent of the opposite sex combined with jealousy of the parent of the same sex (the Oedipal complex); anxiety arising from the anticipation of castration; formation of unconscious defenses against the anxiety, namely, repression, projection, displacement, and symbolization.

Little Hans could much more easily avoid horses than what they symbolized: his father. Of course he paid for these unconscious maneuvers by developing a "neurotic" symptom: a phobia. That is how Freud explained not only Hans' horse phobia but *all* phobias.

How did Freud explain phobias occurring later in life? Do men fear castration forever? What about women? Do *they* fear castration? Freud answered yes to the first and waffled on the latter. Castration fears are unconscious; the person is not aware he has them. In fact, he invests considerable psychic energy to avoid being aware he has them. Simmering in the unconscious mind, castration anxiety can erupt at any time in the form of a phobia, when stress forces a person to "regress" to "an infantile level of functioning" (regression being a way to escape the anxieties of adult life).

If the psychoanalyst could help a phobic patient bring his castration fears into conscious awareness, the fears would be recognized as absurd and would go away. When an adult first realizes he has a secret fear of being castrated, an emotional storm called an "abreaction" may occur, but it passes quickly and the patient's symptoms disappear (sometimes). "Insight" alone does the trick. This was how psychoanalysts proposed to treat phobias in the early days of the movement. Later, insight was downplayed as analysts realized that phobias were not being cured at the expected rate.

As for women, who, so to speak, were born castrated, penis envy becomes the unconscious counterpart of castration anxiety. The implication is that envy and anxiety have similar effects. Many disagree. The penis-envy theory has been much challenged, especially by women.

In any case, most psychoanalysts later concluded that castration anxiety and penis envy were not the only source of phobias. Other unconscious fears also expressed themselves in a symbolic manner as phobias. School phobia might develop from the unconscious fear of killing one's mother. An unconscious desire to exhibit oneself may lead to a fear of crowds. Open streets may assume various symbolic meanings: an opportunity for sexual adventure or the idea that some other person (usually a parent or sibling) might die while one is away from home. Street traffic may symbolize parental intercourse. Woman may develop a fear of shopping because shopping symbolizes taking the mother's place beside the father. Remaining unconscious, repressed, *unnamed*, these unacceptable impulses cause anxiety, the nameless fear.

Sometimes the defense against anxiety is to do what is feared—to the limit! The person afraid of heights becomes a skyjumper; the person with a cat phobia becomes a lion tamer. They now have what are called *counterphobias*. Behaviorists believe fears go away when confronted, but the Freudians say no, the *real* fear is not of heights or cats but of castration (or other unpleasant consequences derived from unacceptable impulses), and the real fear is not being confronted.

As possible sources of anxiety expanded, the possible symbolic expressions of anxiety became increasingly varied. Without abandoning Freud's original idea that phobias represented a "defense against anxiety," latter-day Freudians speculate that even adult nonsexual sources of anxiety may produce phobias. For neoFreudians, such as Alfred Adler and Harry Stack Sullivan, anxiety may derive from a need for power or security. The theories, however, all share a central premise: the person must remain unaware of the source of his or her anxiety. "The fears we know are of not knowing," W. H. Auden said, and a recurrent psychoanalytic theme is that people would rather fear something specific and definite than experience "free-floating anxiety," where they have no idea of the source of their fear. From this point of view the mind can be viewed as a factory forever converting the raw material of free-floating anxiety into concrete fears that can be dealt with, even if the fears are absurd, such as fear of open streets, tomcats, or dining in public.

Freudian theory, and its many variations, resist scientific testing. The analysts have never seemed too concerned about this. They listen to patients and select from the patients' remarks those that fit their theory. William James, the American psychologist, pointed out that you can toss a bag of marbles on the floor and by selectively ignoring certain marbles find any pattern you wish. Not only Freudians do this; as we will see with the behaviorists, the practice is widespread.

LITTLE ALBERT was 11 months old, a phenomenally unperturbable infant. You could put him in a room with rats, rabbits, dogs, and other menacing creatures and he would not blink an eye. A loud noise would make him jump and cry, but that was all. Fearless.

It was 1920 and Dr. John B. Watson was interested in fear. Freud's theories were popular at the time but Watson was a dedicated Pavlovian and did not like Freud's theories. He was sure he could produce fear even in a rock like little Albert, and do it without introducing seductive mothers or scissors-toting fathers. He succeeded.

Dr. Watson took advantage of Albert's startle response to loud noises and the principles of conditioning discovered by I. P. Pavlov, the Russian physiologist. He showed Albert a rat and then banged an iron bar behind his head. Albert jumped and crawled away from the rat. After several exposures to the rat-bang combination, the bang wasn't needed.

Just seeing a rat made Albert cry and race for the opposite side of the crib. Albert had a rat phobia, and John B. Watson was responsible! Then something unexpected happened. Albert showed the same fear response to almost anything furry: a rabbit, fur coat, even a mask of Santa with a beard. Santa had become a *symbol* for the rats. The good behaviorist Watson rejected words like "symbol" but for Freudians it was symbolism, pure and simple, with little Albert "displacing" his newfound fear of rats onto other neutral objects that could then be avoided.

The rat could be avoided too, if Watson did not keep putting it in the crib. He did keep putting it in the crib, however, without banging the bar, and after a while Albert recovered from his rat phobia. Repeated exposure to the phobic object is the road to cure, as Chapter 18 makes clear.

But after Albert was cured, Watson pulled one of his nasty tricks. He brought in the rat and once again banged the bar. He banged it once only but it was enough to make Albert terrified of rats again.

Phobias have this pecularity: they may go away but sometimes return, often for no apparent reason. You think you have the patient cured and there he is, a year later, avoiding elevators again. It is as if, once a phobia has developed, it can be dislodged, but the possibility of relapse persists indefinitely. A sudden loud noise—a brief panic in the supermarket line—and the phobia is in business again. Treating phobias is notoriously frustrating.

Behaviorism is a branch of learning theory. Learning theory holds that fears are conditioned responses and thus learned, just as Pavlov's dog "learned" to salivate when he heard a bell (see Chapter 5). The responses occur if the responder is rewarded but not if the responder suffers (B. F. Skinner's contribution to learning theory). What, then, maintains a phobia? What is the reward?

The reward is powerful indeed. Every time a person avoids a phobic situation, he escapes anxiety. Avoidances may be a nuisance, but they are clearly preferable to panic.

Learning theory is simpler than Freudian theory. Freud also believed phobias were learned, but by a tortuous, convoluted route. A conditioned response is comparatively straightforward. For example, a loud noise or other stressful stimulus produces an *unconditioned* response: anxiety. Pair the stressful stimulus repeatedly with a harmless object, person, or situation and eventually the latter produces a *conditioned* response: anxiety. If, to avoid fear, the person avoids the harmless object, person, or situation, the anxiety becomes a phobia.

Fear can be conditioned to almost anything. In the case of chronically anxious people, behaviorists suspect that perceptions of their own bodies or patterns of light and shade become conditioned stimuli eliciting anxiety. Whenever *anything* happens to us, *something* is going on simultaneously, if only our heartbeat and the ambient temperature. If we are in danger and feel justified fear, our heartbeat and the ambient temperature, by their quite coincidental, simultaneous presence at the time of danger, may later produce *unjustified* fear, or phobia.

Learning theory is not only simpler than Freudian theory but also provides more grounds for therapeutic optimism. One of its basic tenets is that whatever is learned can be unlearned. Habits can be broken, conditioned responses extinguished. The secret of unlearning a fear is to expose oneself repeatedly to the fear situation and eventually, if no harm comes, the fear will go away. Freudian analysts are less optimistic. Fears go back to childhood; they lurk deep in the unconscious mind and have many disguises; it may take years to root them out, if ever.

Freud took a pessimistic view of the treatment of phobias. He himself had a train phobia that persisted even after his self-analysis. He agreed with the behaviorists that the ultimate cure for phobia was to confront the phobic situation and keep it up until the phobia disappeared. As for himself, he stayed off trains as much as possible.

BUT THE BEHAVIORIST theory of phobias also has flaws, the main one being that everybody is exposed to stressful stimuli. Why don't all of us develop phobias? As the behaviorists say, the opportunities for conditioning are limitless. Our very bodies can become a conditioned stimulus eliciting fear, not to mention movies, elevators, and the neighbor's cat. Only a minority of people actually develop phobias—although perhaps a larger minority than was once suspected.

Psychological theories about anxiety leave much unexplained. So, for that matter, do biological theories, but the biologists are at least more united about what they know and don't know.

PART II

A NEW ERA

Every little yielding to anxiety is a step away from the natural heart of man.

Japanese Proverb

7

The Invention of Antianxiety Drugs

> A desire to take medicine is, perhaps, the great feature which distinguishes man from other animals.
>
> Sir William Osler

How To Find a Drug

Dyes have an important place in medical discovery. Bacteriology got its start with the discovery of dyes that stain bacteria so they can be seen under a microscope. The great bacteriologist Paul Ehrlich thought that if dyes could stain germs perhaps they could also *kill* them. This inspired hunch ultimately led to the first sulfa drugs and launched the age of antibiotics.

If anxiety—another scourge of mankind—ever yields to drugs as magnificently as did infectious diseases, this too will be traced in part to dyes. The first specific antianxiety drugs resulted from a search in the early 1930s by a young Polish chemist for a better dye.

His name was Leo Sternbach. A post-doctoral assistant at the University of Cracow, he had been told to find new dyes. In his search he came across some chemicals synthesized in 1891 by two Germans. They were called heptoxdiazines and were believed to have a seven-membered ring (the first mistake). Sternbach—who, like many first-rate chemists, had an aesthetic turn of mind—found them "rather attractive." Among other virtues, they could be modified with great ease. He modified and modified, but produced not a single dye.

Years passed and Sternbach immigrated to the United States. In 1954 he was working in Nutley, New Jersey, as a senior chemist for the Roche Laboratories. He was given a formidable assignment: find a new tranquilizer!

What were called tranquilizers—an unfortunate name—were sweeping the country. Miltown* had just been introduced and was having big sales. The French had produced a drug called Thorazine which, its enthusiasts predicted, would empty the state hospitals. Drug companies were scrambling to improve on these products or come up with new ones. Roche opted for a new one. "Find it," Sternbach was told.

Where there is a profit there is a will and sometimes a way, but how exactly does one "find" a drug? Sternbach had three courses of action:

1. He could tinker with an already existing drug. For example, barbiturates prevent seizures but also produce sedation. Modificaton of the barbiturate molecule led to Dilantin, an anticonvulsant drug that doesn't sedate. Sternbach could have tried the same thing, with a twist: convert a barbiturate from a sedative drug to a tranquilizer (the distinction will be explained later). He decided otherwise. The competitors were already in that game.

2. A second possibility would be to have a theory about the cause of anxiety and construct a drug to fit the theory. That was how sulfa drugs were developed, the theory being that, if dyes stain, they may also destroy. The trouble was that in the mid-1950s nobody had the slightest idea of the chemical basis of anxiety. There were no good theories; Sternbach had to proceed in the dark.

3. Proceeding in the dark, in the drug business, is known as screening. You start by purifying a compound. It may be of natural origins— an extract from barks or berries or roots—or manmade. You give it to animals—usually mice, cats and monkeys—and see what happens. If what happens suggests that the drug might relieve anxiety (for example) in people, you give it to people (assuming it seems reasonably safe). If it works, fine; the company's stock goes up and there are promotions all around.

Screening is an expensive and tedious business. For every 1,000 compounds tested, the chances that one will be effective are very slight. Moreover, even when a drug is effective, it is often not effective in the way intended. Sternbach and every other chemist who works for a drug

---

*Drugs in this book will be called by their trade names rather than their class or generic names. Memorability is the reason. The trade name Librium is highly memorable. It was intended to be; talented people are paid large salaries to produce memorable trade names. Librium *liberates*; Valium makes a person *valiant*; etc. The generic name for Librium is chlordiazepoxide. Most doctors never mastered it on the grounds that the trade name was sufficient. Now that the patent on Librium has expired, doctors may save their patients money by ordering the drug by its generic name.

company must spend a good deal of time praying that the drugs they develop will at least work for *something*.

This chance factor in science is called serendipity. Until Leo Sternbach began looking for a better tranquilizer, all drugs for psychiatric symptoms had been discovered by chance, by serendipity. Thorazine was synthesized in the late nineteenth century in a search for aniline dyes (dyes again!). Resurrected in 1950, it was used as an antihistamine, then as an anesthetic agent, and finally as a drug for nausea before it was finally recognized as effective for psychosis. Miltown resulted from a search for antibiotics. Iproniazid, an antidepressant drug, was first marketed for tuberculosis.

Leo Sternbach went after a better tranquilizer and found one. He probably would have settled happily for a better wart-remover, if he had found that instead.

The Drug That Almost Wasn't

It was a close call. When Roche asked for a new and better tranquilizer, Sternbach remembered his youthful explorations for dye in Cracow. The heptoxdiazines were still attractive, although not the stuff for dyes. They had been ignored since the early 1930s. This was good; *terra cognita* poses no challenges for the explorer. They were easy to make. They could be modified endlessly. And although no member of the group was known to be biologically active, they *looked* biologically active (whatever that means to a chemist) if properly modifed. (Properly modified usually means fantastically lucky.)

For example, Sternbach mused, what would happen if one hooked on a basic side chain? A base, in chemistry, is any molecule or ion that combines with hydrogen. Basic side chains sometimes confer biological activity.

So Sternbach hooked on a basic side chain. Then another, and another. One by one, he had them tested in animals. Nothing happened. They were as inert as marble.

All in all, he cooked up more than 40 compounds, and when none affected animals whatsoever, he gave up (shades of Cracow!).

He remembers the exact day. It was in April 1957. "The laboratory benches were covered with dishes containing crystalline samples, and flasks and beakers with mother liquors which were expected to yield crystalline products. The bench working area had shrunk to almost zero, and a major clean-up operation was in order. Therefore, many old mother liquors which had refused to yield crystalline products were discarded, crystalline products were collected and discarded . . ."

Then Fate stepped in. An assistant pointed to a powder that *hadn't* been tested. An oversight, too busy, whatever. The chemical analysis had been completed: the powder was a base and pure. It had been given one of those James Bond-sounding labels—RO 5-0690—and set aside.

Without hope, Sternbach sent it over to the people who tested chemicals in animals. "We thought that the expected negative result would complete our work with this series of compounds and yield at least some publishable material. Little did we know that this was the start of a program which would keep us busy for many years."

Eureka!

Dr. Lowell Randall was Director of Pharmacological Research at Roche. RO 5-0690 came to him identified only as "white crystalline powder, water soluble as the hydrochloride [salt]." He gave it to mice and cats, the usual procedure.

Two months later, on July 26, 1957, Dr. Randall wrote these historic words: "The substance has hypnotic, sedative . . . effects in mice similar to meprobamate. In cats it is about twice as potent in causing muscle relaxation and ten times as potent in blocking the flexor reflex."

What was meprobamate? Meprobamate was Miltown, the competition. Had Sternbach at last succeeded?

The fact that RO 5-0690 was hypnotic (caused sleep) and sedating was not so important; barbiturates were sedative-hypnotics and *not* the drugs Roche was trying to surpass. The muscle relaxation part was interesting and possibly important; there is a good market for muscle relaxants. However, the startling finding was the ability of this white powder to abolish the flexor reflex. The flexor reflex allows mice to right themselves (and therefore is also called the righting reflex). Why was this important?

When Carl Berger synthesized Miltown in 1950, he learned that the drug blocked the righting reflex. It also did something more interesting: it tamed animals without putting them to sleep or even making them drowsy. Dosed with Miltown, the wildest tigers became pussy cats. Drugs

THE INVENTION OF ANTIANXIETY DRUGS   55

with this taming effect came to be called "tranquilizers."* Meprobamate blocked the righting reflex in mice *before* it had any other effect; this became the acid test for a tranquilizer.

By this standard, RO 5-0690 was a tranquilizer—a tranquilizer ten times more potent than Miltown. It was a big day for Roche. The company later named the drug Librium.

But Does It Work in People?

The story of Librium is a tale of mistakes. The first mistakes were chemical. Sternbach thought he was dealing with seven-member-ring compounds that actually only had six members (maybe not much difference to the customer but a lot to the chemist). Then it turned out that RO 5-0690 differed from Sternbach's earlier inert compounds. It was an entirely new compound, formed by an unexpected molecular rearrangement. By March 1958, the structure was correctly identified as 1, 4-benzodiazepine. Thus, its molecular structure was not identified until nearly a year after Dr. Randall uncovered its pharmacological activity. Not only that: the new compound had a seven-member ring, the number mistakenly attributed to the substance Sternbach had studied in his youth. Finally, the basic side chain was not essential at all. You could unhook it from the molecule and the drug still worked.

So goes the topsy-turvy world of chemistry.

Librium was the first benzodiazepine known to man and quite possibly the first in nature. "Benzodiazepine" became the class name for all the Librium-type drugs to follow.

It's scary to produce a drug that does not exist in nature and then give it to people. But it must be done. Only to a very limited degree can one predict the effect of a chemical on people from its effect on mice and cats.

And here another mistake occurred. This one was almost fatal for the drug.

In early 1958, Librium was administered in largish doses to geriatric patients. The results were discouraging indeed. The drug made the old

---

*Tranquilizer is a confusing word. "Antianxiety drug" is more precise. "Anxylotic" is the word most often used by doctors: the drugs dissolve—"lyze"—anxiety. Thorazine-type drugs are also sometimes called tranquilizers. They dissolve hallucinations more than they tranquilize.

folks stagger, slur their speech, and put them to sleep. It acted, in short, like alcohol or barbiturates, drugs the Roche people were not interested in copying.

As Irvin Cohen, a psychiatrist who early tested the drug, said, it was a "classic example of what happens when a wrong dose is given to the wrong type of patient." It is now known that old people are exquisitely sensitive to Librium and the other benzodiazepines. The wise doctor today gives his geriatric patients benzodiazepines in the dose range for children, and still expects some patients to become addled.

Number One

Somebody at Roche had enough sense not to give up. Psychiatrists were asked to test the drug on patients commonly seen in office practice: "neurotics" whose main symptom was anxiety. It worked. The anxious people became less anxious. There were few side effects.

The medical profession first learned about Librium in the March 12, 1960, issue of the *Journal of the American Medical Association* in a brief note by Dr. Titus Harris of the University of Texas in Galveston. The drug was simultaneously released by the Food and Drug Administration. Within an astonishing short period, Librium and its successors captured a large share of the prescription market. During the 1960s more prescriptions were written for Librium than for any other drug. It was replaced as number one in 1969 by its more potent cousin, Valium, also a benzodiazepine synthesized by Leo Sternbach for Roche. Today, Librium and Valium are sold around the world under more than 1,000 trade names.

Meanwhile, the search for novel, superior, patentable benzodiazepines became intense. Roche has tested nearly 2,000 and many more have been tested by other drug companies. More than 30 are now on the market in the United States and Western Europe, and more are being introduced every year. Most are for anxiety, some for insomnia. Some of the compounds are more potent than others (smaller doses are required) and the duration of action varies. Otherwise the actions are similar, leading to ferocious marketing competition. The potential profits are enormous. For Roche and some other companies, Sternbach's dyestuff has been unmistakably green in hue.

Keys

Before explaining why benzodiazepines opened a new era in the treatment of anxiety, let us examine other drugs available before 1960 that had, among other things, anxiety-relieving properties.

Of course there was alcohol. Alcohol relieves anxiety but, paradoxically, also causes anxiety. About one in 12 users of alcohol becomes addicted. Heavy users are likely to develop liver disease or other serious medical problems. The drug smells on the breath. It puts on weight. The levels needed to reduce anxiety produce obvious intoxication. If a drug company tried to get FDA approval to market alcohol for anxiety, it would probably be turned down.

The same is true of marihuana. Marihuana relieves anxiety but the user is fairly stoned before it happens. It also speeds up the heart and this *makes* some people anxious. It shortens attention span and impairs memory. It is bad for coordination. There is evidence that marijuana may contribute as much to highway fatalities as alcohol.

Morphine relieves anxiety but is addictive. Chloral hydrate and the bromides relieve anxiety but are also addictive. It is incredible that bromides are still available in across-the-counter preparations for "nerves" and headaches. Bromides cause nervousness and headaches as well as relieve them.

Barbiturates were introduced around the turn of the century for epilepsy. They were effective and fairly safe. This is probably because the first barbiturates were long-acting drugs, such as phenobarbital. (Long-acting sedatives tend to be less addictive than short-acting sedatives.) In the 1930s shorter acting barbiturates came on the market, such as Nembutal, and their addictive potential became apparent. Users required increasingly higher amounts to receive the original effect (sedation or sleep) and afterwards experienced severe withdrawal symptoms, including convulsions. These barbiturates were marketed for sleep but, unless the dose was increased, their effects wore off in a short time. Like alcohol, they relieved anxiety but at levels that produced intoxication (although no bad breath). They were lethal in relatively low doses and became the drugs most commonly used for suicide.

Thorazine-type drugs (such as Stelazine and Haldol) are sometimes called "major tranquilizers," but are not particularly effective in treating anxiety. These drugs mainly are used for relieving hallucinations and delusions.

Miltown was introduced in 1950 as the first "true" tranquilizer. If a tranquilizer is a drug that tames aggressive animals without putting them to sleep, Miltown fits the description. Unfortunately, Miltown also produces euphoria and has been abused by many people. It is rarely prescribed today.

How do benzodiazepines compare with the drugs listed above? In general, they have the following actions:

1. They relieve anxiety without necessarily reducing alertness, impairing coordination, or interfering with normal thinking processes. They produce little or no euphoria (in regularly prescribed doses).
2. They relax muscles. They are effective in relieving spasticity from strokes and back strain. The dose required for these problems may cause sleepiness and incoordination. Given this drawback, benzodiazepines are still as good for muscle relaxation as any other drug and probably better than most.
3. They prevent epileptic seizures. Whether they are better anticonvulsants than older drugs such as phenobarbital is arguable, but, given intravenously, they abort status epilepticus—prolonged seizures that may be fatal—better than the other drugs.
4. Benzodiazepines are the safest, most effective drugs for treating alcohol withdrawal and also agitated states produced by many street drugs, such as LSD.
5. Benzodiazepines are the safest drugs in psychiatry. They can be given to belligerent drunks in an emergency room without being lethal (benzodiazepines and alcohol "don't go together," as the saying goes, but the combination is far safer, say, than alcohol and barbiturates).
6. One of the newer benzodiazepines, Xanax, may relieve depression and panic attacks as well as anxiety. The evidence for this is still not conclusive.
7. Benzodiazepines are the best drugs for insomnia. Whether, like barbiturates, they lose their effectiveness in a short time is not clear, but the shorter-acting benzodiazepine hypnotics, such as Halcion and Restoril, leave few if any after-effects the next morning. There is some evidence that when people stop taking them they experience increased insomnia for a few days, but this effect has not been fully evaluated at this writing.

These drugs have one other tremendous advantage over the barbiturates and other hypnotics of the past. It is almost impossible to commit suicide by taking an overdose of them. Very rarely, deaths are reported from overdose, but almost always only when accompanied by other sedative agents or alcohol. Insomnia is a common complaint in depression and depressed people are often suicidal. The virtual suicide-proof aspect of the benzodiazepines is of great value. Benzodiazepines are not perfect. Some people *do* complain of drowsiness, although not many. Some people complain of incoordination or dizziness; but again not many. A very small percentage of patients who take benzodiazepines abuse them, meaning they increase the dose over time and have serious withdrawal symptoms when they stop taking the medicine. Given the billions of prescriptions written for these drugs, the abusers are infinitesimal in number.

There is no substance consumed by humans that some people do not abuse, and undoubtedly more people abuse benzodiazepines, say, than those who abuse water. But whether more abuse benzodiazepines than abuse catsup, milk, or chocolate bars is an open question. The abuse potential has been greatly exaggerated in the press. The book *I'm Dancing as Fast as I Can*, later made into a movie, described tortures-of-the-damned withdrawal symptoms from Valium. High doses of benzodiazepines will indeed produce withdrawal symptoms, including convulsions. There is even evidence that regularly prescribed doses, taken over long periods, result in mild withdrawal symptoms—principally anxiety—when stopped. The problem in evaluating the relationship of withdrawl anxiety to use of anxiety-relieving drugs is that users presumably were anxious before they used the drug, explaining why they took it. On stopping the drug, they simply may be returning to their former anxious state. On the other hand, there may indeed be "rebound" anxiety in some people who use the drug, stop using it, and feel more anxious than they did before. These questions are still difficult to answer.

Another problem, of course, with antianxiety drugs is that everyone agrees that courage and willpower are superior to pills when it comes to coping with anxiety. Everyone also agrees that insulin produced by the pancreas is preferable to insulin administered by a needle. Nevertheless, diabetics die if they don't receive the less perfect form of insulin, and many people experience anxiety inordinately and sometimes to a disabling degree. Some people may be "born" more anxious than others. In any case, some of the anxiety disorders, discussed later in this book, run in families and appear to have a genetically determined physiological basis. Willpower comes easier for some people than others and genes may be the reason.

Better benzodiazepines are on the horizon. "Better" means more selective. Antianxiety drugs should do nothing but relieve anxiety; muscle relaxants should do nothing but relax muscles.

One reason that better drugs are probably inevitable is that Leo Sternbach's approach to finding a new drug is becoming somewhat obsolete. Screening huge numbers of products with a hope of finding a new drug has become prohibitively expensive in an age when the FDA requires stupendous evidence for safety and effectiveness. No longer can a drug move from laboratory bench to the druggist's shelf in a matter of two or three years, as was the case with Librium and Valium.

But the main reason why better drugs are ahead is this: when Sternbach started looking for a new tranquilizer, there were no good theories about the chemical basis for anxiety. Today, there are. The 25 years since Sternbach's discovery have been a period of enormous growth in our understanding of the brain and the biology of emotion.

In the 1950s Sternbach and others produced chemical keys for which there were no known chemical locks. The drugs seemed to work, but no one knew how. People are now learning how; locks are being found.

The new keys will be made specifically to fit them.

When Is a Drug Safe?

It is risky—indeed foolhardy—to suggest that any drug may be "reasonably safe."

When a new drug is first introduced it is often greeted with great enthusiasm. The enthusiasm, just as often, proves short-lived. "Make haste and use all new remedies before they lose their effectiveness," recommended the nineteenth-century physician Sir William W. Gull.

New drugs are also frequently introduced as safer than older drugs. Predicting safety has ruined more reputations than predicting healing powers. When Bayer introduced heroin late in the nineteenth century, the name "heroin" was chosen because the drug was believed safer than morphine, hence "heroic." Of all the words used to describe heroin today, heroic is not one of them. The manufacturers of Demerol, fifty years later, were so convincing in proclaiming their drug safe and nonaddictive that thousands of doctors and nurses proceeded to prove them wrong by becoming addicted. Thalidomide was one more safe new sleeping pill until it was found to produce babies without arms and legs.

Before declaring a drug reasonably safe one should probably wait fifty years or longer to see what really happens after large numbers of people have used it. No amount of testing in animals can assure that a drug will not cause brain rot, cancer, or birth defects in some people. Since only a few users may suffer these bad effects, they may not be detected for years after the drug was taken (just as Parkinson's disease resulting from the 1918 flu epidemic did not start appearing until 20 years later).

Finally, whereas the first drug in a class may be safe, changing the molecule by the addition of a single atom may render it unsafe. Librium, which has been around for 25 years, may indeed be reasonably safe, but what about the benzodiazepines just coming on the market? Human beings are the ultimate test-animals for drugs. Only after a drug has been taken by many people over a long time span can it truly be said to be safe.

Moreover, with drugs that affect the mind, there is always concern about abuse. According to Nathan Eddy, "There is scarcely any agent which can be taken into the body to which some individuals will not get a reaction satisfactory or pleasurable to them, persuading them to continue its use even to the point of abuse." There is no evidence that benzodiazepines are widely abused, but many people have decided they are. A recent example was the decision by the United Nations Commission on Narcotics Drugs to follow a World Health Organization recommendation and declare benzodiazepines to be "controlled" substances. "Controlled" means they require a prescription (which was true in the United States already) and that record keeping be required by the manufacturer and pharmacists. Substances are placed under control when believed to be subject to abuse. The World Health Organization has decided that benzodiazepines have the "capacity to produce a state of dependence and central nervous system depression resulting in disturbances in motor function, behavior, and mood."

The WHO also decided that benzodiazepines were "being abused" to such an extent "as to constitute a public health and social problem."

The available evidence, on the contrary, suggests that the drugs are remarkably unabused, given the huge number of prescriptions written for them. They can indeed produce a "state of dependence," disturbances of motor function, etc., but rarely in the amounts usually prescribed and taken.

In truth, the WHO recommendation was based more on hunch than on data. The hunch may prove accurate. Meanwhile, should the drugs be prescribed if the anxiety is severe or disabling? Most physicians say yes.

One thing is certain: they should not be prescribed to *un*anxious individuals. Quoting Osler: "One of the first duties of the physician is to educate the masses not to take medicine."

8

Does the Brain Make its own Drugs?

> Chemical action must of course accompany mental activity, but little is known of its exact nature.
>
> William James
> *Principles of Psychology* (1890)

The Society for Neuroscience—a focal point for much brain research in America—was founded in 1971. About 250 scientists showed up for the meeting. In 1984, at the 14th annual meeting in Boston, there were nearly 10,000 professionals hearing more than 5,000 presentations on brain research.

The success of the Society in part could be attributed to neurotransmitters: the molecules that nerve cells use to communicate with each other. In 1971 neuroscientists could identify six neurotransmitters. By 1984 they could identify close to 100.

The progress in understanding brain chemistry has been phenomenal. This chapter focuses on three developments relevant to anxiety:

1. The discovery of receptors in the brain with specific affinities for specific drugs, quickly followed by the discovery of natural chemicals in the brain *resembling* these drugs.
2. The discovery of methods for directly visualizing the chemical action which, according to William James, "must accompany mental activity."
3. The emergence of biochemical theories of anxiety derived from neurotransmitter research.

First, however, a few words about the brain and a short vocabulary lesson.

Briefly, the Brain

The adult human brain weighs about two and a half pounds. It is the organ of reason, emotion, and perception. Remove parts of the brain and reason is removed—or emotion, or the ability to see, hear, or feel, depending on the part removed. Crammed into a space the size of a quart jar are ten billion nerve cells, all more or less interconnected. The brain is the most complicated organ in the body.

What is the mind? It is beyond definition. "I think, therefore I am," Descartes said. What does this mean? Psychobabble, seventeenth-century style. If, by mind, we mean consciousness, and consciousness means the ability to respond to certain kinds of stimuli, such as a word or smile, then the mind seems dependent on the brain. Anyway, you can knock out a prizefighter and he doesn't seem to respond to a word or a smile. When he comes to, he says he can't remember people speaking or smiling. If you hit him on the toe, he remembers these things. Apparently mind is not in the toe.

The mind is something the brain *does*. Scientifically, they are inseparable. To say that one thing is psychological (worry, for example) and another biological (cancer, say) is wrong, scientifically. Psychology is a branch of biology. Bile is made by the liver, urine by the kidney, and thoughts by the brain. It's all biological.

A better way of thinking is this: we start out in life and then things happen to us. We are born and we learn and the learning changes us. This is true of everything, including inanimate objects. Shoes. You buy a new pair of shoes and you wear them. As they learn to fit your feet, the shoes change. They keep changing until they wear out, and then the shoe you throw away is very different from the shoe you bought. Brains are the same.

Innate and *learned*: a far better distinction than mind and body, or psychological versus biological. There is another distinction of course: body and soul. It is perhaps the most interesting distinction of all, but not the subject of this book.

Until recently, not much was known about the brain. Too much goes on in too small a space. Our knowledge has come in two stages. First was the wiring stage. Around the turn of the century anatomists began exploring the "machinery" of the brain. The explorations were based mainly on what you could see with the naked eye or a light microscope, or by stimulating or destroying parts of the brain to see what would happen.

In a few decades the wiring of the nervous system was pretty well understood. It consists mainly of pathways and centers. The pathways run up the spinal cord to the brain, carrying messages from skin, muscles, and tendons. Other messages come from our eyes, ears, and other special sense organs. Together, they tell us all we can know about the outer world. Before computers came along, the brain was compared to a telephone switchboard. Messages from the outer world were shunted here and there according to the adaptive needs of the organism. Various clusters of brain cells have particular functions. There is a cluster for speech, another for hearing, a ridge of clusters for outgoing messages to muscles and glands.

Once the wiring was known, progress was slow. Then, in the mid-1960s, new information began flooding the journals. It began, as always, with new technology: the development of microchemical techniques.

What the telescope was to astronomy and the microscope was to bacteriology, microchemistry was to brain research. Events happen on a terrifically small scale in the brain. To know what happens you need measurements in picograms, not micrograms. (A microgram is one-millionth of a gram; a picogram is one-trillionth.) To know, you need to get inside single cells, inside their components, not record the roar of many cells, which is what the electroencephalograph (EEG) does. You want to see what is going on—and you *now* can, in full color, if you have a cyclotron to make radioactive sugar and a computer to plot the radioactivity.

Gas chromatography, mass spectroscopy, radioimmunoassays, nuclear magnetic resonance, the PET scan: all recent, all capable of measuring chemical and electrical events unimaginably small in scale.

The revolution has just begun. To follow it one needs a few terms:

- *Neuron.* A nerve cell.
- *Synapse.* A space between nerve cells.
- *Neurotransmitter.* A chemical released by one nerve cell that crosses a synapse and modulates the firing of an adjacent nerve cell.
- *Receptor.* A protein molecule on the surface of cells that captures the incoming transmitter (or other chemical) and binds it. Each transmitter

has its own unique receptor into which it fits as snugly as a key into a lock.
- *Agonist.* A drug or other chemical that stimulates activity in particular nerve cells.
- *Antagonist.* A drug or other chemical that blocks the action of an agonist.
- *Ligand.* Any chemical that binds specifically with a fraction of biological tissue. Narrowly defined, a ligand is a naturally occurring chemical in the body that has the same effect as a drug. *Example:* endorphin is the ligand for morphine.

The Brain Makes Morphine!

No other area of neuroscience research is as active as investigations of substances occurring naturally in the body called endorphins and enkephalins. To highlight this intense endorphilia, the November 7, 1978, issue of *Science* appeared with a second E missing on its front page. The explanation caption stated that due to a temporary shortage of Es caused by the large number of papers on endorphins and enkephalins, the final E had to be omitted. This bountiful harvest has continued and shows no signs of abating.

Endorphins and enkepalins are small protein molecules discovered in the early 1970s. They took the biological sciences by storm. First, they looked and acted like morphine—a plant substance—but were produced by the body. They suppressed pain. There were indications that they might be involved in a variety of mental disorders, including schizophrenia, depression, and anxiety states. They seemed to be involved in immunity. Acupuncture raised endorphin levels; maybe that was how acupuncture worked! The list of endorphins and of their possible functions grew rapidly.

The discovery of endorphins and enkepalins was made possible by another important discovery. It was found that morphine and morphine-like drugs (called opiates and opioids) attached themselves to specific proteins in the brain called receptors. The opiates were keys. The receptors were locks.

More about locks and keys later. The story is incomplete without a brief introduction to opium.

OPIUM COMES FROM the Greek word *opion* meaning poppy juice. The drug is present in the milky exudate obtained from the unripe seedpod of the poppy *Papaver somniferum*. Opium has been used as a drug since classical Greek times. It deadens pain and causes euphoria. These two properties make opium one of the great mixed blessings of medicine.

In 1803 a German pharmacist isolated an opium alkaloid that he named morphine, after Morpheus, the Greek god of dreams. The use of pure morphine rather than crude opium preparations spread widely. Morphine was more addictive than opium. It was first used in wartime in the U.S. Civil War, sometimes injected into muscle or vein (the hypodermic needle had just been introduced). Morphine addiction became a social problem.

Heroin was introduced in the 1890s as a nonaddictive form of morphine. It turned out to be more addictive than morphine. Since then, pharmaceutical companies have been indefatigable in their search for a nonaddictive painkiller. The search has resulted in a profusion of painkillers (some not yet on the market) whose addictiveness remains yet to be determined.

The long search for a nonaddictive opiate was primarily responsible for the eventual discovery of opiate receptors and endorphins.

THE DISCOVERY OF opiate receptors was made in 1973 simultaneously in three laboratories—two in the United States, one in Sweden. Their existence had been predicted long before. Morphine and other opiates almost *look* like keys that would fit locks, or receptors. They have a T-shaped structure with two broad surfaces at right angles to each other, a hydroxyl group capable of hydrogen bonding and a positively charged nitrogen atom that can form an ionic bond, all suggesting "interactions with a geometrically and chemically complementary receptor site" (to quote one chemist).

Three other points suggested the existence of receptors:

1. Opiates produce highly selective effects at very low concentrations. One opiate, etorphine, is nearly 10,000 times more potent than morphine. It relieves pain and causes euphoria in a dose smaller than one millionth of a gram, making it more powerful than LSD, often cited as the most potent mind-altering substance. For a drug to act in such small doses, the existence of highly specific receptor sites would seem inevitable.

2. Opiates exist in two forms known as optical isomers. Optical isomers are mirror-image molecules with the same chemical composition but, like left and right hands, they cannot be superimposed on each other physically. They can be distinguished by the direction they turn a plane of polarized light. Only the isomer that rotates the plane to the *left* relieves pain and produces euphoria. Only a very smart receptor, presumably, could recognize the "handedness" of the opiate molecule.

3. A very slight modification of opiate molecules produces opiate *antagonists*, substances that block the painkilling and euphoric actions of opiates. Naloxone—an opiate antagonist—instantly reverses the coma produced by morphine overdose. Such a rapid effect suggests that the antagonist displaces the opiate at a highly specific site.

Starting with this knowledge and new technology for measuring small events in the brain, chemists in several countries set their sights on the opiate receptor. First they synthesized a series of morphine derivatives. By slightly modifying the molecules they could heighten or suppress various properties—for example, produce a compound with greater analgesic power and less of a tendency to cause a side effect such as constipation. By such modifications they visualized what the physical form of the opiate receptor might be like.

One problem they faced was that opiates bind to almost any biological tissue or even to the sides of a glass filter. Of the total binding of opiates to brain cell membranes, only 2% of the total represents binding to receptors. Solomon Snyder and Candace Pert at Johns Hopkins found a way around the impasse. They made a labeled opiate so highly radioactive it could be applied in very low concentrations. Thus, receptors with a strong affinity for the opiate would bind with it before nonspecific binding took place. They then washed the tissue rapidly to remove molecules bound nonspecifically. By working with the active and inactive optical isomers and antagonists, they succeeded finally in identifying a specific opiate receptor. To confirm the finding they tested many opiates of varying strengths to determine whether their affinity for the receptor paralleled potency. It did. Potent opiates had a much greater affinity for the receptor than weak ones.

Not every piece of equipment in a modern laboratory is new-fangled. For example, a good way to test the strength of opiate drugs is to use a technique almost as old as pharmacology. It consists of stretching a strip of guinea pig intestine in a test tube. A charge of electricity causes the strip to contract. Opiates inhibit the contraction. The potency of various opiates can be measured by their effectiveness in inhibiting contractions. Opiate antagonists inhibit the inhibition, providing a way to measure the strength of antagonists. It is a lovely system and can be set up at any high school lab for a small investment.

Now for the final chapter in the receptor story: where were the receptors located? Since opiates suppress pain, the brain structures involved in the perception of pain were natural suspects.

There are two brain pathways involved in the perception of pain. Sharp, localized pain passes along one pathway; dull, diffuse pain passes along the other. The first kind of pain is poorly relieved by opiates; dull, diffuse pain is effectively relieved. Sure enough, opiate receptors were richly distributed along the dull pain pathway and relatively absent from the sharp pain pathway.

Opiate receptors were also found in parts of the brain that mediate emotion. Not associated with pain perception per se, they seem involved with the euphoric effects of opiates.

Receptors were also found in the spinal cord, providing evidence that opiate analgesia is mediated in the spinal cord as well as in the brain. Opiates suppress the cough reflex and, sure enough, the nucleus in the brain that mediates the cough reflex is rich in opiate receptors.

Opiates reduce gastric secretion and the brain centers controlling gastric secretion contain many opiate receptors. Opiate receptors are also found in the gut and on the cell membranes of lymphocytes in the blood. Since lymphocytes are involved in immune processes, it suggests that pain, emotion, and even immunity may all share certain common elements.

What are opiate receptors doing in the body? It seems unlikely that such highly specific receptors should have been developed by nature simply to interact with substances from the opium poppy. On the contrary, long before receptors were identified, Claude Bernard and other pioneers in pharmacology predicted the body had drug receptors that were really receptors for counterparts made by the body (ligands). Once the opiate receptor was found, the race to discover its ligand became intense. In 1974, just a year after the discovery of opiate receptors, opiate ligands were found in the brain.

John Hughes and Hans Kosterlitz of the University of Aberdeen found them first. Again, the guinea pig intestine was the hero. As you will recall, electrically induced contractions of the guinea pig intestine are suppressed by opiates. Hughes and Kosterlitz discovered that extracts of brain tissue mimic the effect—and opiate antagonists block the effect! All that was required after that was to isolate the morphine-like factor from the brain of pigs, which they did. They found two molecules, both containing five amino acids. They called the molecules enkephalin, from the Greek for "in the head." One enkephalin they called Met-enkephalin, the other Leu-enkephalin. (If Met and Leu sound like a comedy team, maybe it's because Hughes and Kosterlitz sound a bit like one, too.)

Enkephalins were found hovering near synapses. Opiate receptors were also lined up along synapses. The brain not only made morphine-like substances but placed them near opiate receptors in precisely the parts of the brain where pain and emotional responses are mediated.

Enkephalins looked for all the world like neurotransmitters (and probably are).

Almost simultaneously, Avram Goldstein at Stanford found a morphine-like substance in the pituitary gland that he called beta-endorphin. Endorphins have since become the generic name for all morphine-like substances in the brain, including enkephalins.

A number of endorphins have been identified, purified, and synthesized. At least three different types of opiate receptors have been identified. Before opiate receptors had been found, William Martin at the University of Kentucky predicted there would be three of them, and, at last count, that's what there are.

Not only does the brain make morphine-like substances, they are also found throughout the gastrointestinal tract (which also contains opiate receptors). Why should the brain and the gut be the main site of endorphins? Both rise from the same layer in the developing embryo and perhaps that is the reason. For the millions of victims of irritable bowel syndrome, the gut is preeminently an organ of emotional expression—and maybe this has some relevance. The fact is: nobody knows. The whole receptor-endorphin story is just beginning. More questions are unanswered than answered, but the work goes on, and the future looks hopeful.

Does the Brain Make Valium?

Librium, Valium, and other benzodiazepines are the most selective antianxiety drugs known. They do not exist in nature. They were synthesized, almost by chance, in commercial laboratories searching for a superior tranquilizer.

Some of the benzodiazepines are effective in minute doses, only a few milligrams taken orally. Their concentration in the brain is in the nanomolar range (a concentration of about one benzodiazepine molecule for every 50 billion water molecules). How can such a small amount of a drug possibly affect brain function?

After the experience with morphine—the discovery of opiate receptors and morphine-like substances in the brain—the search for a benzodiazepine receptor was inevitable. The search was successful. Benzodiazepine receptors were found in the brain and nowhere else. Moreover, they only bind with benzodiazepines that change mood. They do not bind with inactive benzodiazepines or with other classes of psychoactive drugs.

The question had been asked: "Why are morphine receptors in the brain if morphine grows only in plants?" The answer was that morphine-like substances do not grow only in plants; they are also manufactured by the brain and very likely modulate responses to pain as well as contribute to our emotional state.

The same question was asked of Valium receptors. If the brain does not make Valium (or other benzodiazepines), why does the brain contain receptors that interact *only* with benzodiazepines?

So far, scientists have failed to find a Valium-like chemical in the brain. However, they have discovered chemicals in the urine of almost equal interest. These chemicals, called beta-carbolines, are antagonists to the benzodiazepines. They compete with benzodiazepines for the benzodiazepine receptor. Occupying these receptors, the antagonists prevent Librium and Valium from having their usual effect. Beta-carbolines, for example, will keep a person awake who has taken a benzodiazepine sleeping pill such as Dalmane. Beta-carbolines also produce signs of anxiety in monkeys and panic in humans. This anxiety is promptly alleviated by administration of benzodiazepines.

But just as most opiate antagonists are not pure antagonists, but mixtures of agonists and antagonists, the beta-carbolines seem to relieve anxiety as well as produce it. They both stimulate the receptor and block it. Thus, benzodiazepines could relieve anxiety because they mimic built-in anxiety-*relieving* compounds or because they block built-in anxiety-*producing* compounds.

Is the brain "wired" for anxiety or calmness? The answer may be both.

Just as there are pain pathways in the nervous system, there may also be anxiety (and calmness) pathways. Some forms of severe anxiety and panic may indeed be due to a chemical imbalance in these pathways. If so, administration of benzodiazepines represents a rational and effective way of restoring balance.

Several types of benzodiazepine receptors vary in their ability to bind different benzodiazepine derivatives. For a time it appeared that the ligand for Valium might be gamma-aminobutyric acid (GABA). GABA is the major inhibitory transmitter in the brain. Benzodiazepines enhance the effects of GABA. This enhancement now appears more important for the anticonvulsant effects of benzodiazepines than for the antianxiety effects.

Opiates, barbiturates, and alcohol also reduce anxiety. Do these drugs also act on benzodiazepine receptors? Clearly, opiates do not. They have their own receptors. The actions of barbiturates and alcohol may, however, be related to benzodiazepine mechanisms. They share the ability of benzodiazepines to enhance the actions of GABA.

Studies of benzodiazepine receptors have just begun, but they have already yielded important information about the neurochemical basis of anxiety. The possibility that the brain manufactures its own Valium raises the possibility that an inborn deficiency of "Valium" might encourage some individuals to self-administer antianxiety agents such as alcohol in an attempt to normalize the brain chemistry.

PET: Seeing Thoughts!

The year 1984 marked the fiftieth anniversary of the discovery of artificial radioactivity. Today we are witnessing the birth of a new technique for the study of the human brain, based on the use of radioactive tracer molecules.

The technique involves an instrument called the PET scanner. PET stands for Positron Emission Tomography.

Here is how it works: A cyclotron produces positron-emitting isotopes (a positron is a subatomic particle of the same mass as an electron). The isotopes are incorporated into natural biochemical substances such as glucose or fatty acids, or drugs, and administered to subjects through inhalation or intravenous (IV) injection.

The positrons are attracted to and collide with their antiparticles (electrons), resembling a "love affair that is doomed from the start" (as one scientist reflected). Both the positrons and electrons are annihilated, but their mass is coverted to energy in the form of a pair of gamma rays that travel in almost precisely opposite directions. The gamma rays arrive simultaneously at crystal detectors arrayed on opposite sides of the PET scanner. A computer then calculates the distribution of the positron-electron collisions and produces images that appear as slices of the brain. From these images, scientists can determine metabolic or chemical changes in brain tissue.

Although positron emission scanning dates back to the 1950s, at least two major steps were necessary before its revolutionary potential could be realized: (1) a computer program for reconstructing an image (first developed for the well-known CAT scan) and (2) the production of positron-emitting isotopes that remain in brain tissue long enough for the scanner to show where the radiolabeled compounds are metabolized in the brain.

IN THE 1970S Barnes Hospital in St. Louis was one of the few medical institutions that possessed a cyclotron. In charge of the cyclotron was Michel Ter-Pogossian, Professor of Nuclear Medicine at Washington University. Ter-Pogossian and his colleagues built the first PET scanner.

In the fall of 1980 Robert Beck, a Chicago scientist, visited Ter-Pogossian's laboratory. Up to that time only monkeys had been PET scanned. Beck volunteered to be the first human to be scanned. The resulting images produced a surprise: an asymmetric distribution of blood in the left temporal lobe of Beck's brain.

"I suspected I had a brain tumor," Beck recalls. Conventional tests revealed no evidence of a brain tumor, but, still concerned, he discussed the finding with a colleague in Chicago. She asked him what he had been thinking about when the scans were being performed.

Beck replied that he had been counting. He had heard a strange clicking noise from somewhere in the machine and had wanted to determine its frequency so Ter-Pogossian could track down its source. The colleague pointed out that Beck was right-handed and that activity such as counting occurs in the left temporal lobe. Beck concluded that "the increase of blood flow in that region of my brain apparently was activated by my mental process."

"Chemical action must of course accompany mental activity," William James had said in 1890, and now, for the first time in history, a man's thoughts could be viewed on a screen and photographed!

(As for the strange noise in the St. Louis machine, Beck says, "I still don't know what the problem was.")

The opportunity to directly visualize mental activity in the brain offers research possibilities of incalculable importance. For one thing, it has greatly improved the ability to study receptors and neurotransmitters and their role in anxiety. It is now predicted that every major university medical center will have a cyclotron and PET scanner within the next five to ten years. The patient who comes to the emergency room with an anxiety attack not only can talk about his symptoms but be provided with a snapshot, in full color, showing where the anxiety has originated in the brain. Meanwhile, with PET scanners and other high technology, investigators can be pinpointing the precise chemical basis for anxiety attacks. With this, almost surely, better treatments will emerge.

Biochemical Theories of Anxiety

Meanwhile, using other techniques, investigators have already started to home in on the chemical basis for anxiety.

It began in the 1960s when Ferris Pitts, Jr., and James McClure at Washington University injected a natural substance called lactic acid into patients subject to repeated anxiety attacks. (Their condition, then known as anxiety neurosis, is now called panic disorder.) The lactic acid produced anxiety attacks in patients with panic disorder but not in volunteers without panic disorder. Since then, hundreds of panic disorder patients have been administered lactic acid, almost always provoking anxiety attacks. Lactic acid administration has virtually attained the status of a diagnostic test. When there is question whether a patient has panic disorder, lactic acid can be given to confirm the diagnosis.

Why lactic acid produces anxiety attacks *only* in patients with panic disorder is not known. Pitts and McClure did their original studies because of evidence that panic-disorder patients had high levels of lactic acid in the blood after exercise. Lactic acid administration has the effect of lowering calcium in the blood and Pitts and McClure speculated that subnormal levels of calcium might be responsible for the attacks. They administered calcium at the same time they gave the lactic acid and, sure enough, it seemed to reduce the intensity of the attacks. However, they then gave a drug that reduces available calcium in the blood and this alone did not produce attacks. The role of calcium in panic disorder remains controversial, but the fact that lactic acid is apparently related to panic disorder in some causal sense has now been well established.

CHAPTER 3 DISCUSSED the James-Lange theory of anxiety which holds that anxiety does not cause increased heart rate, sweating, and respiration but just the reverse: threats to the individual that produce increased heart rate, sweating, and respiration are *followed* by feelings of anxiety. Walter Cannon, a student and friend of William James, disagreed. Anxiety, he believed, began in the brain and the increased autonomic activity, preparing the individual for fight or flight, came instantly afterwards.

There is now evidence that anxiety does indeed originate in the brain. Specifically, it may originate in a tiny cluster of brain cells located in the pons portion of the brain stem. The cluster has a bluish color and therefore is called the locus ceruleus (ceruleus is Latin for blue.) It has recently been found that norepinephrine, one of the brain's most important neurotransmitters, mostly originates from cells in this tiny blue nucleus. The fibers from these cells extend long distances into areas of the brain associated with emotion and up into the cerebral cortex.

If there *is* such a thing as an anxiety center, the blue nucleus may be it. Electrical stimulation of the nucleus in animals produces responses resembling human anxiety attacks. Surgical destruction of the nucleus renders animals incapable of demonstrating fear. Drugs that cause cells in the locus ceruleus to fire provoke anxiety reactions in human volunteers, and drugs that inhibit the firing block spontaneous panic attacks in patients. (Yohimbine is an example of the first type of drug. Valium, some antidepressants, and clonidine are examples of drugs that decrease locus ceruleus firing.) It is interesting that clonidine also blocks withdrawal effects from opiates, some of which resemble panic attacks.

Currently, the locus ceruleus hypothesis is the only comprehensive biochemical theory of anxiety. The locus may be the primary generator of panic or it may cause panic by overstimulation from other biological systems. As Columbia University psychiatrist Donald Klein put it, "Although conflicting data make it premature to accept the hypothesis uncritically, it has stimulated a great deal of research into the anxiety disorders."

Drugs of the Future

Most drugs are "dirty," meaning they have other actions than the ones they are intended to produce. Librium and Valium are not only antianxiety drugs; they also prevent convulsions, relax muscles, and produce undesirable side effects, including sedation and, rarely, physical dependence. The goal of drug companies is to develop drugs that have only *one* action—the action sought.

There is now a furious race in the drug industry to produce an antianxiety drug that does nothing but reduce anxiety. The key to the search, almost everyone agrees, is the receptor.

Since the discovery of opiate receptors in 1973 and benzodiazepine receptors a few years later, developing new drugs no longer rests on chance. It is now guided by the idea that all processes in the body—taste, touch, pain, fear, responses of the body's immune defenses against disease—are probably mediated by receptors. Receptors regulate the manufacture of proteins; the permeability of cell membranes (controlling what goes in and out of cells); and the shape, movement, and growth of cells. Ligands, antibodies, hormones, and neurotransmitters involved in passing messages between nerves exert their effects by binding to receptors. This is even true of foreign invaders of the body. Bacteria and viruses rely on receptors, too. These receptors, indeed, allow the body's immune defenses to mark them as foreign and consequently to act on them. An antibody recognizes a foreign receptor (called an antigen) and binds to it.

Once again: Think of a receptor as a lock to a door. In order to bind to a receptor, a ligand, a hormone, antibody, or drug must possess a molecular structure that acts as a key to that door.

Hundreds of receptors have been identified; many more undoubtedly will be found. The job for a drug designer is to find the most important receptor involved in the physiologic reaction that he wants to control. Next he determines its structure. Finally, he designs a drug to fit it.

He actually has two choices. He can design a drug that stimulates a particular physiological response: an agonist. Or he can develop a drug that blocks a response: an antagonist. Think of an agonist as a drug that unlocks a door like a key. Antagonists gum up the lock, preventing the key from opening it. Vaccines, for example, work by mimicking locks (antigens) and by stimulating the body to make lots of keys (antibodies) to fit them.

Having decided which receptor interests him, the drug designer then unravels its structure in order to make a drug to fit it. Receptors are proteins. They may contain as many as 300 amino-acid building blocks that are linked, folded, and twisted to form a complex three-dimensional shape.

There are several ways to determine the shape of a receptor. If you are lucky, you can actually look at it with an electron microscope or bombard it with X rays (another way of looking at it). You may read the DNA recipe of the gene that controls the amino-acid sequence and then try to work out how that string of amino-acids would fold up to form the protein. Computers are immensely helpful here.

Computers, in fact, make possible recognition of receptors in general. Computers help decide whether one protein resembles another because the two behave in the same way. You may know the key but not the lock; if so, the "picture" of the key can be dropped on a grid, drawn on a computer screen, and deformed into a three-dimensional shape. The result is a sort of molecular cast or mold for the receptor.

Receptors are crucial in determining the shape of keys. Three-dimensional models of receptors and of proposed drugs can be spun together on a computer terminal and fitted together like jigsaw puzzles.

Shape is not everything. The way in which receptor and ligand react to water and the distribution of electrical charge in them may matter. A good physical fit might be strengthened by a sort of electrical glue if your receptor has a small negative charge and your drug a small positive charge. Again, computers are indispensable.

Once the designer knows what he wants, it is then up to the chemist to make it. These days, chemists can put together almost any molecule from materials off the shelf if they want to, and can do so relatively quickly and cheaply. They have become adept at modifying selected sites on a molecule, without altering its other properties. Indeed, just as they have learned to modify genes, chemists are learning how to modify proteins so as to change their properties in a controlled way. The technique for doing this is called protein engineering.

The revolution in drug development beyond doubt will result in medications that relieve anxiety and do nothing else. There will be no side effects. They will not cause cancer or birth defects. They will be nonaddictive, meaning more and more of the drug will not be needed to produce the same effect (tolerance) and withdrawal symptoms will not occur when the drug is stopped (physical dependence).

People will become dependent on them—psychologically dependent—just as they are on their wives and automobiles. Is this so bad? By supplying chemical courage, they may discourage natural courage, or what we call coping skills. They may render entire populations so serene that no one protests, demonstrates, marches in streets, writes letters to editors: fair game for dictators. This was what happened 50 years ago in *Brave New World*, when Aldous Huxley, with amazing prescience, fantasized a drug called soma which tranquilized and did nothing else: exactly what pharmaceutical companies around the world are searching for—and will almost certainly find.

Do we really want to live in a brave new world? Most Huxley readers say no. Give us anxiety.

True, but remember: some people are born more anxious than others. Some are incapacitated by anxiety. If the brain indeed makes Valium, perhaps their brains don't make enough. Others experience pain more than others; maybe they were born with an endorphin deficiency.

The deficiencies will soon be corrected in safer, better ways than they are now. Should we deny painkillers to those in pain? A better Valium for those wracked by anxiety? Insulin for the diabetic?

Remember, the physical courage, the tolerance to pain, so much admired in football players may be less a product of willpower and character than of a super supply of endorphin. The easygoing person at the next desk may be blessed with large amounts of built-in Valium (while the guy next to him, born with too little, has to pop them). Just as noses and feet come in different sizes, brain chemistry varies from person to person.

Maybe we shouldn't be too judgmental. Maybe we should view a better Valium as a blessing for some and an irrelevancy for most.

Anyway, it's coming.

9

The New Classification

Diagnosis precedes treatment.

Russell John Howard

Everyone experiences anxiety on some occasion, in some degree, but some people have incapacitating anxiety. They have sudden panic attacks for no apparent reason; they are afraid to leave the house, fly in airplanes, or eat in public; they wash their hands twenty, fifty, one hundred times a day to hold terror at bay.

These people have anxiety disorders. In the past five years these disorders have been reclassified with a precision and clarity unknown in the past, and this has led to increasingly specific treatments for specific disorders. The "shotgun" approach to treating anxiety—offering the same pill or couch or group therapy for every anxious person—is becoming a thing of the past.

This chapter explains how the new classification came about, beginning with the question "What is disease?" Many people, including some physicians, balk at viewing anxiety disorders as diseases. Without some understanding of the *historical* concept of disease, the rest of the story makes little sense.

Psychoanalysis dominated American psychiatry for 40 years. With the decline of psychoanalysis, starting in the mid-1960s, a new psychiatry emerged that was more scientific, more like other medical specialties, and better equipped to sort out the sufferers of disabling anxiety into separate groups that could be studied and helped.

First, comes a discussion of disease, then a historical note on the domination of psychoanalysis, and finally the events leading to a new classification of the anxiety disorders.

Is Nostalgia a Disease?

Nostalgia *was* a disease, at least briefly. Two hundred years ago a famous physician named Leopold Auenbrugger described the disease as follows:

> When young men, not yet arrived at their full growth, are forcibly impressed into a military service, and thereby at once lose all hope of returning safe and sound to their beloved homes and country, they become sad, silent, listless, solitary, musing, and full of sighs and moans, and fianlly quite regardless of, and indifferent to, all the cares and duties of life. From this state of mental disorder nothing can rouse them—neither argument, nor promises, nor the dread of punishment; and the body gradually pines and wastes away.
>
> This is the disease nostalgia.

The disease was not confined to morbid thoughts and feelings. Auenbrugger was a pulmonary specialist. He had discovered percussion, which involves tapping the chest with a finger to detect congestion. He tapped the chests of the young men with nostalgia and found congestion. On autopsy the lungs were "firmly united to the pleura, and the lobes were callous . . . and purulent." (Some of his cases of nostalgia would probably be called tuberculosis today.)

Nostalgia didn't make it as a disease. The history of medicine is a graveyard of diseases that came into prominence for short periods and then expired. Among them, historian Robert Hudson lists lardaceous liver, gleet, chlorsis, febricula, and status thymicus: all bona fide diseases in times past and now meaningless words to most physicians.

A "disease" of particular interest to psychiatry was masturbatory insanity. It was a common diagnosis in the nineteenth century, filling asylums with its victims. The "masturbatory personality" was clearly spelled out in Tuke's *Dictionary of Psychological Medicine* (1892):

> The face becomes pale and pasty, and the eye lustreless. The man loses all spontaneity and cheerfulness, all manliness and self-reliance. He cannot look you in the face because he is haunted by the consciousness of a dirty secret which he must always conceal. . . . He shuns society, has no intimate friends, does not dare to marry. . . . Later he becomes a liar, a coward and a sneak . . . and finally he sinks into melancholic dementia, relieved only by occasional excitement due to a temporary revival of his jaded passions.

Three cures were recommended. One consisted of dried beetles, which, paradoxically, were also used as an aphrodisiac. Castration in men and removal of the clitoris in women were much practiced, although there was a period when placing a silver ring around the foreskin was considered the treatment of choice for men.

Masturbatory insanity had a longer career than nostalgia, but finally disappeared in the twentieth century. The *coup de grace* was Alfred Kinsey's report in 1948 that 98% of men masturbated. Masturbatory insanity had been common but not *that* common.

Perhaps one reason there have been so many diseases is that few physicians have bothered to define disease. Here is a more or less modern definition.

A disease is a cluster of symptoms and/or signs with a more or less predictable course. Symptoms are what patients tell you; signs are what you see. The cluster may be associated with physical abnormality or may not. *The essential point is that it results in consultation with a physician who specializes in recognizing, preventing, and, sometimes, curing diseases.*

It is hard for many people to think of anxiety disorders and other psychiatric problems as diseases. For one thing, psychiatric problems usually consist of symptoms—complaints about thoughts and feelings—or behavior disturbing to others. Rarely are there signs—a fever, a rash. Almost never are there laboratory tests to confirm the diagnosis. What people say changes from time to time, as does behavior. It is usually harder to agree about symptoms than about signs. But whatever the psychiatric problems are, they have this in common with "real" diseases—they result in consultation with a physician and are associated with pain, suffering, disability, and death.

Another objection to the use of terms like disease and disorder in psychiatry arises from a misconception about disease. Disease often is equated with physical abnormality. In fact, a disease is a category used by physicians just as "apples" is a category used by grocers. It is a useful category if precise and if the encompassed phenomena are stable over time. Diseases are conventions and may not "fit" anything in nature at all. Through the centuries, diseases have come and gone, some more useful than others, and there is no guarantee that our present "diseases"—medical or psychiatric—will represent the same clusters of symptoms and signs a hundred years from now that they do today. On the contrary, as more is learned, more useful clusters surely will emerge.

Sometimes society decides what constitutes a disease. Suicide and homosexuality are examples. Both were sins in the eighteenth century, crimes in the nineteenth, and diseases in the twentieth, although there was dissent from the majority view in all three centuries. (Today, most psychiatrists view suicide as a symptom of disease rather than a disease.) Progression from sin to crime to disease is not unusual. Alcoholism went through a similar metamorphosis. Hudson said, "If a medical and social consensus defined freckles as a disease, this benign and often winsome skin condition would become a disease. Patients would consult physicians complaining of freckles, physicians would diagnose and treat freckles, and presumably, in time we would have a National Institute of Freckle Research."

The diseases that last are useful categories. They possess two qualities to a high degree. First, they are reliable. This means that two or more physicians, observing the same signs and symptoms, should arrive at the same diagnosis. Second, they are valid. In medicine, this usually means the disease predicts something: what will happen if the person is not treated, the complications of the disease, whether the disease is contagious, which treatments work.

The anxiety disorders described in the last part of this book are, by and large, useful categories. They are reliable: most health practitioners can agree on the diagnosis. They also have predictive power: something is known about their natural histories and which treatments are most effective. They are relatively unambiguous and precise, and have enhanced communication between doctors and patients and helped investigators to learn more about the nature of the conditions.

And yet this new classification is only a few years old. Psychiatry has lagged behind the rest of medicine by at least a half-century in putting together a useful classification of diseases. In America, psychoanalysis was largely responsible for this.

Psychoanalysis in the Saddle

"Verily, the stone which the builders refused has become the headstone of the corner."—Psalms 118:22

"The Emperor not only has no clothes, he has bad skin."—Ambroise Bierce

"There is some truth in psychoanalysis, as there was in Mesmerism and in phrenology (e.g., the concept of localization of function in the brain). But, considered in its entirety, psychoanalysis won't do. It is an end-product, moreover, like a dinosaur or a zeppelin; no better theory can ever be erected on its ruins, which will remain forever one of the saddest and strangest of all landmarks in the history of twentieth-century thought."—Peter Medewar

Sigmund Freud once said that he worried more about the future of psychoanalysis in the United States than elsewhere because it was received with "open arms and too uncritically."

"Open arms" puts it mildly. For over 40 years, beginning in the mid-1930s, Freudian psychoanalysis dominated American psychiatry. Almost all the chairmen and professors of psychiatry were psychoanalysts or at least enthusiastic about Freudian theory. Generations of American medical students were taught psychoanalytic theory as received truth. Articles by analysts appeared regularly in American medical journals. The New York literary establishment, led by Lionel Trilling and other giants, adopted psychoanalysis with religious fervor. In novels, movies, newspapers, and other media, psychoanalysis and psychiatry were treated as identical specialties.

The takeover of psychiatry by the psychoanalysts occurred only in America. Other countries remained wary, and Russia, already having an ideology, made psychoanalysis a punishable crime. Even in America psychoanalysis faced tremendous opposition before it became the "headstone of the corner."

Before speculating about the reasons for this 40-year infatuation, a few comments are needed about psychoanalysis as a treatment and a theory.

Freud himself expressed a rather low opinion of psychoanalysis as a treatment. Toward the end of his life he remarked that psychoanalysis would be remembered as a psychology of the unconscious and not as a method of treatment. "I do not think," he wrote, "that our successes can compete with those of Lourdes. There are many more people who believe in the miracles of the Blessed Virgin than in the existence of the unconscious." Freud's lack of enthusiasm for psychoanalysis as a treatment may partly explain the reluctance of analysts to check the results of their therapy. If faith is the basis for whatever success they achieve, other methods that also rely on faith presumably would do as well—and Lourdes is cheaper!

Freud defined psychoanalysis as a "theory of the unconscious." The idea that much of human behavior is unconsciously motivated is very old and Freud said he could only take credit for discovering a "scientific method" for exploring the unconscious. The fact that he refers to "the" unconscious reveals his major interest, namely, defining the "structure" of the mind (which he subdivided into id, ego and superego). According to his biographer, Ernest Jones, Freud's three main contributions to human knowledge were his discovery of the structure of the mind, the meaning of dreams, and the Oedipal complex.

Compared to his disciples, Freud even seemed skeptical about his theories. Summing up his work, he wrote, "I can say that I have made many beginnings and thrown out many suggestions. Something will come of them in the future. But I cannot tell myself whether it will be much or little."

Nevertheless, Freud constantly referred to his work as scientific. As the word is usually defined, however, it does not apply to psychoanalysis. Scientifically, psychoanalytic theories have a crucial flaw: they cannot be disproved.

Karl Popper, a philosopher whose views about science are accepted by most scientists, said that *"a theory which is not refutable by any conceivable event is nonscientific."* Psychoanalytic theories can explain everything; whatever happens confirms the theory.

Popper illustrated his point with two examples of human behavior: that of a man who pushes a child into the water to drown it, and that of a man who sacrifices his life to save the child.

> Each of the two cases can be explained with equal ease in Freudian and in Adlerian terms. [Alfred Adler was a neoFreudian who retained unconscious motivation as the hub of his theories.] According to *Freud*, the first man suffered from repression . . . of his Oedipus Complex, while the second man had achieved sublimation [of his Oedipus Complex]. According to *Adler*, the first man suffered from feelings of inferiority and needed to prove that he dared to commit some crime. So did the second man (whose need was to prove that he dared to rescue).

Popper continues:

> I could not think of any human behavior which could not be interpreted in terms of either theory. It was precisely this fact—that they *always* fitted, that they were *always* confirmed—which in the eyes of their admirers constituted the strongest argument in favor of these theories. It began to dawn on me that this apparent strength was in fact their weakness.

Popper compared psychoanalysis to Marxism: "A Marxist could not open a newspaper without finding on every page confirming evidence for his interpretation of history; not only in the news, but also in its presentation—which revealed a class bias of the paper—and especially of course in what the paper did *not* say."

The analysts emphasized that their theories were constantly verified by their "clinical observations." Once, Popper wrote Adler about a case "which did not seem particularly Adlerian. Nevertheless, Alder had no difficulty in analyzing the case in terms of his theory of inferiority feelings, although he had not even seen the child. Slightly shocked, I asked how he could be so sure. 'Because of my thousandfold experience,' he replied; whereupon I could not help saying: 'And with this new case, I suppose, your experience has become thousand-and-one-fold.' "

Not only are psychoanalytic theories not subject to the test of "falsifiability" (to use Popper's favorite word), but they also suffer from another defect: they are virtually incapable of predicting anything. Freud never claimed they had predictive value. Psychoanalysis, he said, was a "postdictive science." This, for most scientists, would be a contradiction in terms. A scientific theory is valid to the extent that it predicts events.

By disdaining prediction, the analysts cannot show that their treatments are effective. To know whether a treatment is effective, one must follow a group of patients with similar symptoms over a period of time to see how well they respond, and then, if they seem to do well, compare this particular treatment with another treatment to determine whether the effectiveness is due to the treatment or reflects nonspecific factors. The analysts specialize in single case histories. Scientific hunches may derive from case histories, but effectiveness of treatment rarely can be determined from a single case or even a small group of patients. Individuals with "terminal" cancer sometimes survive. In clinical medicine it is axiomatic that anything can happen to a single patient.

Because of its unscientific nature, psychoanalysis has been famous for producing silly as well as untestable theories. Among them are the following (recently cited by Donald Klein):

1. Melancholia is due to retroflexed rage at the lost object combined with oral fixation and a severe superego.
2. Paranoia is due to latent homosexuality.
3. Obsessive compulsive disorder is due to repressed anal eroticism.
4. Agoraphobia is a defense against repressed prostitutional drives.
5. Menstrual cramps indicate a lack of acceptance of womanhood.

6. Pregnancy morning sickness is due to a repressed fantasty of oral impregnation.

These notions, as Klein says, have undergone "disuse atrophy." Their credibility stemmed from the authority of their originators. As the authority waned, so did the credibility.

To summarize: Psychoanalytic theories, whether propounded by Freud or revisionists such as Adler and Jung, are neither scientific nor supported by data as effective treatments. Moreover, they have produced some of the silliest theories in history, compared to which the phrenological theories of Franz Josef Gall were monuments to science. Why, then, did psychoanalysis dominate almost every aspect of American psychiatry for more than 40 years?

Here are some possible explanations:

1. There is a strong authoritarian tradition in medicine. During the last half of the nineteenth century and first half of the twentieth, the authorities most respected by American physicians were European. In the heyday of bacteriology, hosts of young American physicians with academic aspirations made a pilgrimage to Germany to work with great bacteriologists. In the 1920s and 1930s, psychiatrists and neurologists from the United States went to Vienna to see the great Freud.

If they were lucky, Freud would analyze them. If unlucky, there was always a Viennese disciple to fall back on. Some Americans went to Vienna because they were not having much success with psychoanalysis in their New York or Boston office practices. Obviously there was something wrong with their "technique," which could only be corrected by time and money invested on Professor Freud's couch. Many came home disillusioned; Freud wasn't racking up many successes himself.

In any case, to Americans of that generation, any theory set down in the German language by a Viennese professor had to be a good theory. It never occurred to them to ask for evidence. In medical school they had never asked for evidence when following a professor on rounds and *he* often was an American. In America, Freud had it made.

2. Americans are (or were) natural hero worshippers. Cynical Europeans adulated Lindbergh for a day, but the Americans loved him for a lifetime. The romantic strain in the American character created a hero of Freud as did no other country. Like Grant at Vicksburg, Freud impressed Americans as a tough, indomitable genius who overcame tre-

mendous odds in pursuit of a noble ideal—in the case of Grant, victory for the North, in the case of Freud, *truth*. (Americans, perhaps more than any other people, regard truth as a noble ideal. Nixon would have survived unscathed anywhere else.) Incidentally, both Grant and Freud had impressive beards and both smoked 30 cigars a day for most of their life and both died of cancer of the soft palate. One does not need to be a Freudian to suspect unconscious symbolism at work in the American psyche.

A word about Freud's genius. Even today it is almost impossible to give a lecture on psychoanalysis without paying tribute to Freud's genius. You then may proceed to attack his ideas *in toto*, as long as the genius part is established.

Was Freud really a genius? That depends on one's definition of genius, but Freud's ardent disciple and biographer, Ernest Jones, gave a rather interesting explanation for Freud's reputation as an original thinker in the same league as Copernicus and Darwin. Jones wrote: "Freud has been regarded as a revolutionary genius who introduced novel and disturbing ideas. As a result of my researches I came to the *unexpected* conclusion that hardly *any of Freud's early ideas were completely new* [emphasis added]." Then he gives a long list and adds, "This is a formidable list, which I cannot discuss here in detail, and yet it can be shown that there are broad hints of all of them in the writings of previous workers *with which Freud was thoroughly familiar.*"

Why was this conclusion unexpected by Jones? Obviously because he had not read of them previously. But there is another aspect of the matter. Why had Freud left a faithful, devoted disciple in such ignorance? Freud, in a letter, says that he stopped reading other writers because they expressed his own ideas better than he could. Did he wish to save his disciple Jones from such a deception? Or had he forgotten? If so, what motive had he for forgetting?

Percival Bailey, a neurosurgeon, psychiatrist, and heretic, raised these questions at a meeting of the American Psychiatric Association in 1956. With wonderful irony he concludes: "You will pardon me for asking such a question, but I am told that I must learn to think dynamically—that is, in terms of motives. I leave you to speculate on the answers [why Freud forgot]. I remember only that Freud was a very ambitious man."

In any case, even Freud's detractors concede he was a genius, one reason apparently being that he "forgot" to give credit to other geniuses.

3. Americans are joiners. A secret pining for elitism seems to be partly the reason. One nice thing about being a Rotarian is that most people are not Rotarians.

And one nice thing about being a psychoanalyst is that most people (including most psychiatrists) are not psychoanalysts. In its heyday only about 15% of American psychiatrists were fully trained and full-time analysts. Full training includes a personal analysis. The psychoanalytic institutes admitted fewer than 50% of applicants and many of those accepted ran out of money or patience, or were "unanalyzable," and fell by the wayside. Thus, most psychiatrists were excluded from the club.

True, the analysts say that a man can be a good psychiatrist without having been analyzed but they do not act as though they believe it. For more than 40 years they made each new generation of psychiatrists believe that if they had not been analyzed, they belonged to a lesser breed.

Fortunately, except in some small enclaves on the East and West coast, this belief is no longer held.

4. Freudian theory is basically a theory of unconscious conflict and the American culture is cloven with conflict. Vying for dominance in the American personality is the Puritan versus the frontiersman, the missionary versus the drunken trapper, and, above all, *being* good versus *making* good. Billy Graham may have successfully combined the latter two, but he has rare gifts.

5. Analysts recruited thousands of inexperienced physicians into their ranks during World War II, thanks to the historical accident that a forceful and impressive analyst, William Menninger of Topeka, became a brigadier general in charge of Army psychiatry. The official "training manual" for Army psychiatrists was undisguised Freudian gospel. In the Army, authority is even more respected than in medical school and if a brigadier general who is also a psychiatrist says it is true, it must be true.

6. Americans are always looking for new religions. Ernest Jones, Freud's biographer, avowed that Freud's theories were "more likely than any other to yield the secrets that were perplexing me, about the nature of the soul, the purpose of life and the means of controlling our animal nature." From the ranks of American psychiatry one can still hear, faintly in the background, "Amen!"

7. Freudian theory is easy. All the skill needed is to grasp two or three principles and, like dice, roll them to the ground and there is the answer. Also needed is a certain glibness and ability to sound plausible in every situation.

The easiness of Freudian theory indeed has posed a caste and economic problem for psychiatrists practicing psychoanalysis. If any social worker—or longshoreman—can quickly master psychoanalytic theory, why shouldn't patients see *them* instead of paying big bucks to psychiatrists who in turn had to pay big bucks to attend medical school? Freud said that analysts did not need to be physicians, but physicians who are analysts almost universally disagree and one sometimes senses a pecuniary motive.

These are some of the possible reasons Americans embraced psychoanalysis. There are undoubtedly others. But sometime in the mid-1960s the orthodoxy started to crumble, and that is the topic of the next section.

Heresy on the Mississippi

"The follow-up is the great exposer of truth, the rock on which many fine theories are wrecked and upon which better ones can be built."—P. D. Scott

The 1950s were a time of complacency for the Freudians who controlled American psychiatry. The theory had become so entrenched at all levels that it seemed nothing could shake it. Perhaps the first inkling of change appeared in a brief article in the April 27, 1954, issue of *Time* magazine. Most of the Freudians who read it probably shrugged and said "What else is new?"

This article announced that a drug developed in France and tested in Canada had now appeared on the American market and was having a "sensational" effect on the chronically ill psychiatric patients in state hospitals (which accounted for more than half of the hospital beds in the country). The drug was Thorazine, the first of a long series of antipsychotic drugs introduced in the 1950s and 1960s.

The Freudians were probably right to shrug. Since henbane in Socretes' time, many new "wonder" drugs for psychiatric illnesses had come and gone. Greeted with enthusiasm, they never quite lived up to their promise. One of the best was opium, but it was in such disrepute by the mid-twentieth century that it was rarely given to psychotic patients.

But Thorazine was different, and for once *Time* was right: the effects were sensational. Thorazine did not cure schizophrenia—the illness affecting the largest group of state hospital patients—but it reduced and sometimes abolished the hallucinations and delusions that characterize the disease. It also reduced agitation and made the patients *seem* more sane even if they weren't. This had a healthy effect on the hospital environment. As the patients stopped screaming obscenities and defecating in hallways, the ward personnel treated them more and more as human beings. An optimism was born that may have been just as therapeutic as the new drugs. State legislators voted funds for new hospital buildings; the old ones got a paint job and curtains for the windows. As the optimism increased, the patients got better; and as the patients got better, the optimism increased.

The optimism, to be sure, was not a total blessing. First, it was not entirely justified. The drugs were really not that good. Also, a well-intentioned but unrealistic movement to empty the state hospitals swept the land. They were indeed emptied: in the 1960s their occupancy dropped to a third of what it had been ten years before, but the result was heavily drugged schizophrenics released onto the streets where they were unemployable and unwanted—and often went off their medicine and became psychotic again. A nice, clean hospital would have been much more humane.

Nevertheless, Thorazine opened a new era in psychiatry. Drugs for psychoses were followed by drugs for depression and drugs for anxiety. None were panaceas but all were superior to what had been available in the past. A new subspecialty in psychiatry known as psychopharmacology had appeared, and it challenged the dominance of psychoanalysis. For a time, some of the brightest young psychiatrists bridged both worlds, being analyzed while they mastered the new drugs. As time passed, more and more were content just to master the new drugs.

And it was all gaudily reported by the press, with cover stories in news magazines as naively optimistic about the new drugs as they had earlier been about psychoanalysis. The patients and their families read the stories and the word "revolution" sent them off to find a doctor who had a prescription pad rather than a couch.

Another development in the 1960s further loosened the grip of psychoanalysis on psychiatry. President Kennedy got a community mental health bill passed and, almost overnight, small towns throughout the land had their own community mental health centers.

This, too, gave psychiatrists something else to do other than psychoanalysis. Couches were conspicuously absent from community mental health centers. The rage became short-term therapy, not requiring couches, often conducted by social workers, but every community mental health center wanted a psychiatrist even if it couldn't afford one. Being wanted is a good feeling, almost as good as being accepted by a psychoanalytic institute. Anyway, everyone agreed that psychoanalysis had nothing to contribute to community mental health (meaning most mental health).

Disenchantment with the community mental health centers followed this movement by only a few years. Like the drugs, the centers were overtouted. The results were a disappointment to both psychiatrists and the public. Both contributed, however, to a growing skepticism about psychoanalysis in the medical community. In the early 1960s committees formed by medical schools to find new heads for departments of psychiatry started looking for nonanalysts to fill these positions.

With the chairman's power of appointment, his or her view can be crucial. Predictably, the new nonanalyst chairmen appointed nonanalyst faculty. These new faculty members began spreading nonanalytic views to medical students and residents. Analysts still had their voice in departments but it wasn't the only voice. Increasingly, psychiatry departments became known as "eclectic" (a word connoting quackery in the nineteenth century). Representing several schools of thought, they sometimes created a babble of competing views that confused medical students and perhaps contributed to the decline in the number of them who chose psychiatry as a specialty over the past 15 years.

Not all the new departments were eclectic. Some didn't have an analyst on the property. They were as single-mindedly devoted to a particular brand of psychiatry as the most ardent Freudians had been. The departments became known as organic, or biological, or neo-Kraepelinian—all misnomers, as will be explained later.

The first, or at least best known, of these nonanalytic, noneclectic departments was at Washington University in St. Louis. The faculty began gaining a reputation as mavericks in the 1950s while psychoanalysis was still "mainstream" psychiatry. They didn't attack analysis; they simply ignored it. They did studies based on the "medical model," meaning psychiatric problems were conceptualized as diseases wherever possible. This meant an emphasis on diagnosis, on the course of the illness, and on its outcome, or prognosis. To know the course or outcome, follow-up studies were needed—the "rock on which fine theories are wrecked and upon which better ones can be built" (to quote P. D. Scott). Follow-up studies were as highly regarded by the psychiatrists at Washington University as were unconscious motives by the Freudians.

In the early 1960s, Eli Robins, a Harvard-trained psychiatrist and biochemist, became head of the department of psychiatry at Washington University. Robins was an unusual man. He did clinical and epidemiological studies and at the same time carried out experiments in biochemistry. His clinical studies of suicide are classics, and he also collected the brains of suicides to study them biochemically. The department grew rapidly under his direction. Some of the staff did follow-up studies, some biochemistry, and others, like Robins, both.

By the late 1960s, Washington University's psychiatry department was gaining prominence for the research it produced. Its reputation as a maverick was changing. Psychiatrists and neurobiologists from around the world visited the school. Grant funds were abundant. Its senior faculty were offered chairmanships in other universities and some took them. Washington University wasn't becoming more like the rest of psychiatry—the rest of psychiatry was becoming more like Washington University. In 1980, its point of view, while not universally held by any means, became the "official" view of American psychiatry. This happened in a curious way.

One of the frequent visitors to Washington University as its star began to rise was Robert L. Spitzer, Professor of Psychiatry at Columbia University. Spitzer was trained as a psychoanalyst but this did not blind him to other approaches. One approach he liked was Washington University's. It was empirical, practical, and it valued diagnosis.

So did Dr. Spitzer. In fact he was the chairman of a committee the American Psychiatric Association formed to revise its official diagnostic manual, to be called DSM-III (*Diagnostic and Statistical Manual of Mental Disorders*, Third Edition). He took his job seriously. Earlier manuals had been skimpy affairs sprinkled with Freudian jargon. ("A neurosis occurs because of displacement, conversion," etc.). The Freudians had little interest in diagnosis. They exploited their position of power to transform the diagnostic manual into psychoanalytic propaganda. Nobody seemed to mind; the Adlerians and Jungians were just as indifferent to diagnosis.

Spitzer was going to change that. Whereas DSM-I and II had combined clinical description with theories of causation, he was determined to eliminate the theories on the grounds they were unproven. He introduced specific diagnostic criteria—rules for making a diagnosis that enhanced communication and made research more reproducible.

Dr. Spitzer and his committee labored on the book for five years. They used hundreds of consultants. Scores of field trials were held to see if the diagnostic criteria were reliable (they were).

Psychoanalysts were invited to participate but most declined; so not much psychoanalysis got into DSM-III. A lot of Washington University thinking got in, which was predictable. One-third of the DSM-III task force consisted of Washington University-trained psychiatrists. "What could I do?" asked Spitzer. "They were the only ones interested in diagnosis."

The final product came out in 1980. It was 494 pages long compared to 119 pages in DSM-II. It had some 240 categories, perhaps too many. Spitzer's position was to include everything on the grounds that you can always eliminate categories but can't add them after the book is published. This seemed fair enough.

DSM-III became an instant best seller and has made a handsome profit for the American Psychiatric Association. In many schools it is used as a textbook. This is unfortunate. It lacks references and every good textbook should give its sources of information. The long list of diagnostic criteria suggests a certainty and consensus that do not exist. The criteria were, after all, invented by people sitting around a table, often by taking a vote. At bottom, they are arbitrary. They are not based on studies except those predigested by the consultants. But to medical students and others, the book is "official." It gives highly specific instructions and medical students like specific instructions.

With DSM-III, psychoanalysis fell from grace. The new descriptive, organic, biological, neoKraepelinian psychiatry, even if offensive to some and rejected in part by almost everybody, was now mainstream.

What to call this psychiatry? Descriptive it is, but "organic" and "biological" are misnomers. They suggest that descriptive psychiatrists give more pills than others (they don't); are opposed to psychotherapy (they aren't); or believe all psychiatric illnesses are caused by biochemical abnormalities (who knows?). All psychiatric illnesses *are* biological if biology is broadly defined to include learning, memory, intelligence, conditioning—in short, all of psychology.

NeoKraepelinian is also a misnomer. Emil Kraepelin was indeed the father of descriptive psychiatry, but he was also a true believer. He attributed psychiatric illnesses to toxins or constitutional defects. In truth, he knew no more about the causes of illnesses than do present day psychiatrists.

Perhaps the best description of the new psychiatry is "agnostic." Its proponents believe that when the cause of a disorder is unknown, it is best to say so and refrain from theorizing unless it produces testable hypotheses.

THE STEPS that led to DSM-III—the new classification—were long and convoluted. To understand them, one must know something about the historical development of the concept of disease; DSM-III, after all, consists of diseases (called disorders, but the two words are synonyms).

A classification of diseases has been slow to emerge in psychiatry, principally because of the 40-year dominance of the field by psychoanalysts. Among the diseases in DSM-III are anxiety disorders. Millions of people suffer from these disorders but only in the last five years have they been separated into categories that permit diagnosis and the development of specific treatments. It has been long in coming.

PART III

THE ANXIETY DISORDERS

> The thing I fear most is fear.
> Michel de Montaigne

If every doctor, in whatever branch of medicine, would consciously and earnestly study pain and fear, both in a scientific and humane regard and with full profit of their literature, and would then set about to improve his understanding and management of cases on the basis of that study, the practice of medicine and the public health would win benefits, less spectacular maybe, but falling little short of those conferred by the discoveries of Pasteur and of Lister.

John A. Ryle
Fears May Be Liars

10

Rules

> Not chaos-like, together crushed and bruised,
> But, as the world harmoniously confused:
> Where order in variety we see,
> And, where, though all things differ, all agree.
>
> Alexander Pope

DSM-III describes five anxiety disorders:

- Generalized anxiety disorder
- Panic disorder
- Phobic disorders
- Obsessive compulsive disorder
- Post-traumatic stress disorder

Phobic disorders are divided into simple phobias, social phobias, and agoraphobia. This book adds a fourth category—school phobias—because of its importance to many families.

There is a chapter for each disorder, with each chapter following a similar format. First the disorder is defined. Then diagnostic criteria are presented. This is followed by an elaboration of the symptoms, with occasional case vignettes to make a point.

Then comes an estimate about the frequency of the condition; *whom* it is most likely to affect; *when* it is most likely to occur; *what happens* to the symptoms over a period of time; and the illnesses with which it might be mistaken. Comments on treatment conclude the chapter.

The diagnostic criteria to some extent are arbitrary. As noted in Chapter 9, they were invented by a committee. Whether they are good criteria depends on whether they bring together homogeneous groups about which accurate predictions can be made: whether the symptoms tend to go away or get worse, how they respond to particular treatments, etc.

This is the test of the value of any classification system, and it can only be applied *after* the fact—after hundreds of patients have been studied and followed over periods of time. Countless studies are currently in progress to test the predictive value of the categories in DSM-III. So far, the results have been encouraging.

The diagnostic criteria are not sacrosanct. For research purposes, they should be meticulously followed; if not, differences in research findings may reflect differences in criteria. However, in evaluating individuals for treatment, they should be viewed as loose approximations, as guidelines.

The criteria consist of *inclusion* criteria and *exclusion* criteria. For example, to fulfill the criteria for generalized anxiety disorder, a person must (A) have symptoms from three of four categories; (B) the symptoms must be continuous for at least one month; and (C) the patient must be at least 18 years of age. These are the inclusion criteria.

The exclusion criterion is that (D) the symptoms must not be due to another disorder such as depression or schizophrenia.

Deciding which diagnosis takes precedence over other diagnoses was not the work of the DSM-III committee or any other, but evolved gradually over many years. It is known as the hierarchical organization of diagnostic classes. It is based on the following rule:

A disorder high in the hierarchy may have features found in disorders lower in the hierarchy, but not the reverse. In making a diagnosis, "higher" disorders preempt "lower" disorders—for example, schizophrenia preempts obsessive compulsive disorder because, while schizophrenics may have obsessions, obsessional patients do not have delusions or hallucinations (symptoms of schizophrenia).

Anxiety is the most common symptom of all. It can be observed in all mental disorders as well as many physical illnesses. Thus, if a patient has symptoms of schizophrenia—hallucinations and delusions—*plus* anxiety symptoms, schizophrenia is the "correct" diagnosis. It is correct because the course of the illness, the disablement, the response to particular treatments will be more characteristic of schizophrenia than of an anxiety disorder.

The same is true of major depressive disorder (formerly called endogenous depression). If a patient is depressed and has classical features of major depressive disorder (e.g., insomnia, suicidal thoughts), *plus* a full range of anxiety symptoms, depression is the "correct" diagnosis. The course is likely to be episodic, the symptoms disabling, suicide a serious possibility, electroshock therapy effective; all typical of depression, less characteristic of anxiety disorders.

Diagnostic dilemmas arise when the clinical syndrome is not classical—when, for example, the anxiety symptoms are as striking or more disabling than the depressive symptoms. Depressive disorders usually preempt anxiety disorders, but not always; "clinical judgment" may preempt the hierarchy! The clinician may fudge, using terms like "depression with anxiety features" or "phobic disorder, with secondary depression." But when circumstances compel a hybrid diagnosis, something is lost. Specifically, the clinician loses much of his ability to predict, and his treatment will be based more on intuition and guesswork than solid information. At these moments the "art of medicine" is no idle phrase.

Mixed clinical pictures also may have another basis: the patient may have two illnesses. The patient, say, may have schizophrenia and, independently, an obsessional disorder. To the extent that both are influenced by heredity, such mixtures should not be unexpected. In order to have children, people must become acquainted. Many become acquainted in doctor's waiting rooms, in group therapy sessions, in hospital wards. When a manic-depressive meets and marries an obsessional, the children of the union may have illnesses that are hard to classify.

Not much is said about the causes of anxiety disorders for the simple reason that the causes are unknown. There are interesting theories, some of which were covered in the first section of this book. But DSM-III is "atheoretical," meaning its purpose was to describe disorders and not parade theories as explanations.

However, patients will often try to explain why they have anxiety attacks or phobias; it is only human to want to explain things, particularly unpleasant things. Often the "cause" is something that happened in childhood. Mother was gone, a storm came up, a scream was heard in the attic. Or something real and serious: a car wreck, the death of a parent, the experience of nearly drowning.

But beware of connections. Millions of people go through car wrecks and bereavement and hear screams in the attic as children and do not later have anxiety attacks or phobias.

If a traumatic experience caused anxiety disorders, most of us would have anxiety disorders, and most of us do not.

It is not that life experiences aren't connected with emotional symptoms. Connections obviously occur and are important. It is just that the connections are much more complicated than we usually realize.

Walter Kaufmann says this about explanations:

> No single explanation can really explain human behavior; it can at most illuminate human behavior and allow us to see something we had not seen. . . . An accident may be considered a paradigm. Why did it happen? The road was icy at that point. And the driver of the small car was in a great hurry because he was late for a crucial appointment, because the person who had promised to pick him up had not come. And his reflexes were slower than usual because he had had hardly any sleep that night because his mother had died the day before. And just before the accident his attention was distracted for one crucial second by a very pretty girl on the side of the road, who reminded him of a girl he had once known.
>
> Yet he might have regained control of his car if only a truck had not come toward him just as he skidded into the left lane. The truck driver might have managed not to hit him, but . . . . If we add that the truck driver had just gone through a red light and was, moreover, going much faster than the legal speed limit, the policeman who witnessed the accident, as well as the court later on, might discount as irrelevant everything said before the three dots and be quite content to explain the accident simply in terms of the truck driver's two violations. *He* caused the accident. But that does not rule out the possibility that the other driver had a strong death wish because his mother had died, or that he punished himself for looking at an attractive girl the way he did so soon after his mother's death, or that the person who had let him down was partly to blame.

To repeat: beware of explanations. They aren't needed to recognize and treat anxiety disorders and rarely give the full picture.

11

Generalized Anxiety Disorder

> The most influential obstacle to freedom of thought and to new ideas is fear, and fear which can with inimitable art disguise itself as caution or sanity or reasoned skepticism or on occasion even as courage.
>
> Wilfred Trotter
> *Collected Papers*

Generalized anxiety disorder is a condition that usually begins in late adolescence or early adulthood and is manifested by persistent feelings of tension, jitteriness, being on edge—as if something terrible may happen at any minute. The person is not in danger. The anxiety is free-floating, meaning it is unattached to anxiety-provoking situations. The person knows not what he fears, but lives in a constant state of dread.

Accompanying the apprehension is the full panoply of physiological fear responses provided by nature for surviving real dangers (described in Chapter 3). The person often knows, intellectually, that he is not in danger but is on the constant lookout for danger. Tense, drawn, jumpy, he has trouble carrying on normal activities and is sometimes hard to live with, but is not incapacitated. Most of the time he gets by. "The mass of men lead lives of quiet desperation," Thoreau said. The individual with generalized anxiety disorder is simply more desperate than most.

Diagnostic Criteria

A. Generalized, persistent anxiety is manifested by symptoms from three of the following four categories:

1. *Motor tension*: shakiness, jitteriness, jumpiness, trembling, tension, muscle aches, fatigability, inability to relax, eyelid twitch, furrowed brow, strained face, fidgeting, restlessness, easy startle

2. *Autonomic hyperactivity:* sweating, heart pounding or racing, cold, clammy hands, dry mouth, dizziness, light-headedness, paresthesias (tingling in hands or feet), upset stomach, hot or cold spells, frequent urination, diarrhea, discomfort in the pit of the stomach, lump in the throat, flushing, pallor, high resting pulse and respiration rate
3. *Apprehensive expectation:* anxiety, worry, fear, rumination, and anticipation of misfortune to self or others
4. *Vigilance and scanning:* hyperattentiveness resulting in distractibility, difficulty in concentrating, insomnia, feeling "on edge," irritability, impatience

B. The anxious mood has been continuous for at least one month.
C. Not due to another mental disorder, such as a depressive disorder or schizophrenia.
D. At least 18 years of age.

Elaboration

The criteria required that a person must be at least 18 years old to qualify for the diagnosis of generalized anxiety disorder. True, most sufferers begin having symptoms in their late teens or early twenties, but there are plenty of exceptions. Judging by his *Collected Letters* and biography by Scott Elledge, E. B. White, the essayist, had a classical case of generalized anxiety disorder and began having symptoms as a young child.

White often said he was born lucky but was also born scared. Of his early anxieties he once gave a fairly specific account:

> As a child I was frightened but not unhappy. I lacked for nothing except confidence. I suffered nothing except the routine terrors of childhood: fear of the dark, fear of the future, fear of the return to school after a summer on a lake in Maine, fear of making an appearance on a platform, fear of a lavatory in the school basement where the slate urinals cascaded, fear that I was unknowing about things I should know about.

White's fears were more than "routine."

> The normal fears and worries of every child were in me developed to a high degree; every day was an awesome prospect. I was uneasy about practically everything: the uncertainty of the future, the dark of the attic, the panoply and discipline of school, the transitoriness of life, the mystery of the church and of God, the frailty of the body, the sadness of afternoon, the shadow of sex, the distant challenge of love and marriage, the far-off

problem of a livelihood. I brooded about them all, lived with them day by day. Being the youngest in a large family, I was usually in a crowd but often felt lonely and removed. I took to writing early, to assuage my uneasiness and collect my thoughts . . .

Many famous writers had abnormal fears in childhood. Many were worriers and hypochondriacs as adults. A good number would probably qualify for the diagnosis of generalized anxiety disorder. Indeed, the vigilance and scanning, the hyperattentiveness, listed above as cardinal features of the disorder may be important in the creative process.

DSM-III says the symptoms must last for at least one month. This requirement was introduced to eliminate the "mass of men" who feel tense and anxious for brief periods but rapidly settle back into their normal lives of "quiet desperation." In truth, generalized anxiety disorder usually lasts for many years. White had symptoms throughout his lifetime. The symptoms often mimic those of physical illnesses and lead to consultations with a variety of medical specialists.

White had one classical symptom—"jelly legs"—not included in the criteria. It is often mistaken for Ménière's disease, a disorder of equilibrium. In a *New Yorker* piece called "My Physical Handicap, Ha, Ha," White writes:

> I will be walking along the street, say, and will take three normal steps in a forward direction: then, as I am about to set my foot down for the fourth step, the pavement moves an inch or two to the right and drops off three-quarters of an inch, and I am not quick enough for it. This results in my jostling somebody on my left, or hitting the corner of the Fred F. French building a glancing blow. It was fun for a few days, but I have recovered from the first fine ecstasy of dizziness, and am getting bored with it. Once I sidled into a police horse, and he gave me back as good as I gave him.

White saw a doctor who "partly by stealth, partly by cunning," gained entrance into his middle ear, hoping to discover there the secret of the dizziness.

> He blew and he blew, setting up Eustacian williwaws of great intensity. . . . Physicians on the trail of a man's dizziness are explorers of a hardy sort. They are ready to go anywhere, on short notice, traveling light. They speak of toxicity, and set out for the Yukon. Yet there are undoubtedly toxic secretions in a man which the medical fraternity know very little of.

The medical fraternity never found the source of White's toxins, although they kept looking throughout his lifetime. Like most sufferers he was frequently afflicted with dizziness and afraid he would pass out, causing great embarrassment. Once, at a White House press conference, he suffered an attack of "nervous stomach and dizziness so severe that he nearly passed out and had to leave the Oval Room in a hurry." He worried constantly about having cancer.

> Last night in the middle of this letter I began to feel queer in the head . . . and when I touched my forehead it felt as soft as a piece of putty. I examined it in the mirror and the whole front of my head was swollen like the breast of a pigeon. Tumor of the brain, I told myself and collapsed on the bed in one of my panics. . . . I prepared for the end . . . and went to bed, full of flatulence, dizziness and fear.

Next morning, a doctor told him he had a sunburn.

Once he thought he had throat cancer. (The sensation of having a lump in the throat is a classical symptom of anxiety disorders.) He saw a doctor who told him his Adam's Apple was "tangled up" with his psyche.

Many people with generalized anxiety disorder are hypochondriacs but most, like White, know, without being told, that anxiety is their basic problem. White compared his anxiety to mice. Instead of bats in the belfry, he said, he had a mouse in his mind. In a poem called "Vermin" he wrote:

> The mouse of Thought infests my head,
> He knows my cupboard and the crumb.
>   Vermin! I despise vermin.
> I have no trap, no skill with traps,
> No bait, no hope, no cheese, no bread—
> I fumble with the task to no avail.
> I've seen him several times lately.
> He is too quick for me.
> I see only his tail.

Earlier in this book it was said that anxiety strikes "from the rear." White could only see the tail. Superb writer that he was, he had many metaphors for his anxiety, once comparing it to a "kite caught in the branches" of his head.

It would be misleading to suggest that mainly sensitive, artistic people have generalized anxiety disorder. As far as is known, it afflicts people of every background, although women may be more susceptible than men. Here is the case history of an edgy electrician.

A 27-year-old, married electrician complained of dizziness, sweating palms, heart palpitations, and ringing of the ears of two year's duration. He also experienced dry throat, periods of uncontrollable shaking, and a constant "edgy" and watchful feeling that often interfered with his ability to concentrate.

Because of these symptoms he had seen a family practitioner, a neurologist, a neurosurgeon, a chiropractor, and an ENT specialist. He had been placed on a hypoglycemic diet, received physiotherapy for a pinched nerve, and told he might have "an inner ear problem"

For the past two years he had had few social contacts because of his nervous symptoms. Although he sometimes had to leave work when the symptoms became intolerable, he continued to hold a steady job. He tended to hide his symptoms from his wife and children, to whom he wanted to appear "perfect," and reported few problems with them as a result of his nervousness.

Generalized anxiety disorder is distinguished from other anxiety disorders mainly by the *absence* of certain symptoms. For the diagnosis to be made, the patient should not have frequent panic attacks, obsessions, or phobias (for reasons described in Chapter 10). At least, these symptoms must not be the dominant symptoms. To receive the diagnosis, the patient should not be a heavy drinker or abuse mood-altering drugs. Both alcohol and drugs such as barbiturates, after their sedating effect, produce anxiety symptoms indistinguishable from those of generalized anxiety disorder. Finally, any person with chronic anxiety should have a physical examination to exclude medical illnesses that cause anxiety. The most common of these illnesses are hypoglycemia and disorders of the thyroid and adrenal glands.

Treatment

People with generalized anxiety disorder rarely need to see a psychiatrist. A psychiatrist may be useful in making the diagnosis and excluding the possibility of other psychiatric conditions, but the treatment of generalized anxiety consists primarily of supportive psychotherapy and medication. There is no evidence that psychological management more extensive than reassurance has any better effect than reassurance alone. Prolonged and expensive forms of psychotherapy are not usually helpful.

The patient does need an understanding physician, however, one who understands the illness and its natural history and is willing to discuss the syndrome with him. Telling the patient that there is nothing wrong physically is usually not enough. Some patients resent the implication that their symptoms might be psychological. Moreover, there is evidence that anxiety disorders are to some extent inherited and may involve a chemical imbalance (see Chapter 8). The physician should agree that something is wrong and should describe the syndrome in lay terms. Many patients are relieved by such an explanation and thereby become receptive to further reassurance.

The physician should also be skilled in the administration of antianxiety drugs. These drugs, described in Chapter 7, are very useful in the control of anxiety but it is important that they be given in the proper dosage and that the physician be aware of side effects and possible dangers.

A wide variety of antianxiety drugs is now available. They differ mainly in their side effects and how long they remain in the body. It *is* possible to become physically dependent on the drugs and their use should be monitored carefully. Usually, it is best to take them on a regular basis for several months and then gradually reduce the dosage. A patient should never abruptly discontinue taking these medications. With gradual reduction of dosage withdrawal symptoms rarely occur. After a few months the patient may find he doesn't need to take the medication except perhaps sporadically. On the other hand, some patients need the medication for long periods but can usually manage with low doses, sometimes taking an extra tablet or capsule half an hour before a particularly stressful occasion. Prescriptions should not be refillable, so the doctor remains in charge of the treatment.

A class of drugs called beta-blockers are also useful in controlling anxiety. Propranolol is the best known of the group. These drugs block the physical manifestations of anxiety, especially palpitations and racing of the heart. The physical manifestations of anxiety are part and parcel of the *feeling* of anxiety. Sometimes blocking the physical manifestations will abolish the feeling.

Beta-blockers, by the way, are more toxic than antianxiety drugs. The physician's control of the dosage is especially critical.

Sometimes people with generalized anxiety disorder become depressed. When this happens it is called a "secondary" depression, but

GENERALIZED ANXIETY DISORDER 107

the treatment is the same as for any depression: antidepressant medication and supportive psychotherapy.

Some authorities, indeed, believe generalized anxiety disorder is a form of depression in which depressive symptoms masquerade as anxiety symptoms. This is a minority view, but nevertheless any patient with generalized anxiety disorder who fails to improve on antianxiety drugs should be given antidepressant medication to see if it helps. It sometimes does—sometimes dramatically.

Supportive psychotherapy—a term introduced above—means listening carefully to the patient's complaints and giving encouragement and advice. It does *not* mean trying to help patients understand why they are anxious. The cause of generalized anxiety disorder is unknown. To suggest otherwise is misleading.

Nevertheless, much of psychotherapy practiced today emphasizes "insight" or "understanding." Since people with generalized anxiety disorder are likely to become involved with such therapy, a few comments about psychotherapy in general are in order. (A few treatment approaches do not fit the following description and are mentioned later.)

There are many schools of psychotherapy, but they all have one thing in common: they involve two or more persons talking to each other, and one of the discussants is supposed to know more than the other about what is going on. He knows more because he has gone to school and received a degree or has had some other special experience that makes him an authority on the subject, such as having been "in treatment" himself.

"Talking it out" or "getting it off your chest" has long been valued as a means of relieving certain forms of distress, such as a smoldering resentment or feelings of bereavement. The amount of relief thus attained, however, is often exaggerated and sometimes, rather than feeling better for having talked about their problems, people feel worse; talking about the problems reminds them of them. At any rate, psychotherapists tend to attribute the success they believe they obtain to unique features of their particular school of psychotherapy.

Each school has a doctrine. Each doctrine, almost without exception, has a founder, a great man whose theories and writings are highly esteemed by his followers, so that schools of psychotherapy inevitably come to resemble religions. The therapist is a votary and the patient a supplicant hoping to overcome his problems by *understanding* them. In psychotherapy, understanding has roughly the same role as grace in religion. There is an assumption in psychotherapy—sometimes denied but almost always present—that understanding the nature and origins of a particular problem helps a person overcome the problem.

This has come to seem almost self-evident, but it is not clear why. A patient with cancer presumably would benefit little from understanding the cause of his condition. What we call understanding, in fact, is simply the point where curiosity rests. There is no area of inquiry where you cannot go a layer deeper if you are able to and so desire. Electricity is understandable as a "flow of electrons" but what is an electron? An electron is an "elementary particle consisting of a charge of negative electricity," according to Webster, which shows how understanding is frequently an exercise in circular reasoning. In psychotherapy understanding occurs when the patient agrees with the therapist about what is wrong with him. When the patient's theories about himself coincide with the therapist's theories, he is suppose to improve. Does he?

There is hardly any *scientific* evidence that "insight" psychotherapy helps anyone. There have been a number of studies, but few have met even the minimal requirements for a scientific study. Yet thousands of people make their living giving psychotherapy and millions have received psychotherapy, many of them feeling they have benefited from it. They may indeed *have* benefited from it; after all, there is no evidence that psychotherapy does not work. But the burden of proof, as always in such matters, is on the proponents.

As mentioned earlier, psychodynamic psychotherapy, derived from the theories of Freud, dominated American psychiatry for some 40 years. The unwillingness of the proponents of this school to check their results is the principal reason so little is known about whether the treatment is useful or not.

A type of psychotherapy called "cognitive therapy" *has* been tested scientifically and found to be moderately useful for depression (particularly when combined with antidepressant medication). This form of therapy concentrates on training patients to substitute positive thoughts for negative thoughts (a cognition is a thought). Patients are encouraged to think of themselves as strong and resilient rather than weak and worthless.

Thought-editing (cognitive therapy) may also be useful in anxiety disorders. Talking to one's self—giving one's self instructions—may be useful if the person is not in the grip of a powerful emotion. Before thought-editing can help a severely anxious person, a drug may be needed to reduce the level of anxiety.

The principle behind thought-editing is that a person can only think one thought at a time. If he dwells on how fearful he is, he cannot simultaneously think of himself as tough and resourceful. Conversely, if he tells himself repeatedly that he is tough and can "take anything," this will supplant the anxious thoughts, assuming the emotion behind them is not too powerful.

Many people go through life with mental bumper stickers that they repeat to themselves from time to time to avoid thinking gloomy or anxious thoughts. "Life is short!" and "To hell with it!" are good bumper stickers. "There is just today and I can get through *anything* for a day" is another. If this sounds suspiciously like "the power of positive thinking," it is unfortunate. Just because a bad book was written about a useful technique doesn't make the technique less useful.

Humor may also be helpful in controlling anxiety—as E. B. White discovered in childhood. Some therapists try to help patients overcome their fears by deliberately exaggerating them. This technique derives from the observation that anxiety may be increased by the endeavor to avoid anxiety. If, for example, a patient is unreasonably afraid of having a heart attack, he may be encouraged to "try as hard as possible" to have one "right on the spot." Called "paradoxical intention," this often introduces an element of humor into the situation, which may itself be helpful.

As noted, controlling the physical symptoms of anxiety may reduce the subjective feeling of anxiety. Here are some ways for reducing these physical symptoms without using drugs.

Anxiety often causes people to overbreathe—to hyperventilate. People with anxiety disorders commonly worry about breathing. "I feel like I am running out of air" is a frequent complaint. This particularly happens in hot, stuffy, crowded places, or rooms without windows.

Concern about breathing causes the person to breathe faster and more deeply. This, in turn, produces *more* physical symptoms: lightheadedness, dizziness, tingling sensations, even fainting. The cause is wholly chemical. The overbreather sucks in too much oxygen and blows out too much carbon dioxide. Carbon dioxide controls the respiratory rate. The lower the carbon dioxide level in the blood, the slower the respiratory rate. When the person becomes conscious of his breathing becoming irregular or slowing down, his anxiety leads to increased hyperventilation. This worsens the chemical imbalance and he may faint because of too little carbon dioxide (whereupon, while unconscious, his breathing will automatically return to normal and he wakes up).

There is an elegantly simple treatment for overbreathing: a paper bag. Closing off the mouth and nostrils with a paper bag over the head, the carbon dioxide is trapped in the bag, signalling the respiratory center in the brain to start firing again. This may be low-tech indeed, but it works and there is no doctor fee.

Anxiety is also accompanied by an awareness of the heart beating. The heart thumps or speeds up or seems to be skipping a beat. Skipped beats—known as premature contractions—are particularly frightening. In his book *The Anxiety Disease*, David Sheehan describes how the heart muscle beats its normal beat, then contracts again too soon. "These two beats are felt as twin beats very close together. Since it has worked double time, the heart now takes a longer break before going back to its normal work rhythm again—hence the awareness of a pause before the next beat." The anxiety that produced the double beat is now compounded by fear of having a heart attack. When, on top of this, the heart starts to race, rare is the patient who isn't convinced the end is near.

Sheehan suggests several ways for slowing down the heart when it starts to race. One is to massage the carotid body. "The carotid body," he explains, "is a lump located on the carotid artery, which carries blood from the heart to the brain; it is found in the neck just below the ear lobe at the level of the angle of the jaw. If you place your finger there, you will feel the artery pulsate. Massaging the artery gently at that point causes a reflex slowing of the heart.

"The carotid body is full of little receptors that check the rate of the heart and the blood pressure; if the speed or pressure get out of bounds, they send a signal to the brain to slow the heart by nerve action. Massaging the carotid body tricks it into sending those 'slow down' messages out in greater numbers." When the heart rate is slower, it is also less likely to produce skips. Never massage both carotid bodies simultaneously; the heart rate may become dangerously slow.

Another way to slow the heart rate is to take advantage of the "dive reflex." The dive reflex is highly developed in fish. "When threatened or in danger," Sheehan writes,

> fish may dive suddenly nose-first to save themselves. The cold water and increased water pressure on their faces stimulates nerve endings in the face

and elicits the dive reflex, which causes their heart rate and rate of metabolism to slow down. Their bodies go into low gear as a mechanism of protection and to conserve energy for fight or flight. Man has a vestige of this reflex: stimulating the face, mouth, or throat with cold water, or putting pressure on the closed eyes, will cause some reflex slowing of the heart rate.

Sheehan's book has some excellent descriptions of people with generalized anxiety disorder, as well as some practical pointers for relieving symptoms. It is recommended.

12

Panic Disorder

> O lift me from the grass!
> I die! I faint! I fail!
> My cheek is cold and white, alas!
> My heart beats loud and fast.
>
> Percy Bysshe Shelley
> *Indian Serenade*

Paul Dudley White, the famous Boston cardiologist who rode a bicycle to keep fit (long before fitness was a national preoccupation) and was Eisenhower's doctor after his heart attack, fairly often was referred patients who had symptoms of heart disease but perfectly normal physical examinations. White reassured them that they did not have heart disease and sent them on their way but, being a curious man, he often wondered what happened to them.

In the late 1940s he decided to find out. With admirable persistence, using newspapers, radio, and whatever means he could find, White managed to locate 173 patients whom he had examined 20 years previously. They all had received a diagnosis of neurocirculatory asthenia, a medical jawbreaker now known as panic disorder. All had come to his office with heart-attack symptoms: chest pain, often extending down the left arm; trouble breathing; pounding and racing of the heart, often with skipped beats; and extreme apprehension. The symptoms came on suddenly, as would a heart attack, and went away almost as abruptly. A physical examination, including electrocardiogram, was normal. The individuals were not unusually apprehensive between attacks, although with each succeeding attack they worried more and more about having heart disease and were often hard to convince that this was not the case.

How were these people 20 years later? The answer is: pretty good.

Most were still living. Few had developed evidence of real heart disease. They had continued to have attacks and some continued to see doctors for reassurance, but their lives, in general, had been normal. They had come to "live with" their "nervous condition." Their life expectancy had not been reduced and they were no more susceptible to physical or other mental disorders than the normal run of people.

They did find that their nervous condition had other names than neurocirculatory asthenia. The most common was anxiety neurosis, introduced by Freud in 1895. Occasionally they were told they had "effort syndrome," based on the observation that the attacks seemed to occur more often when they were physically active. "Irritable heart" was another term applied to their malady. Some received the oldest diagnosis of all: Da Costa's syndrome, Da Costa being a Civil War surgeon who described the condition almost exactly as it is described today—now with the title of panic disorder.

Thus, the American Psychiatric Association's new classification of mental disorders (DSM-III) includes one of the oldest syndromes in psychiatry, with specific diagnostic criteria added for research purposes.

Diagnostic Criteria

A. At least three panic attacks within a three-week period in circumstances other than during marked physical exertion or in a life-threatening situation. The attacks are not precipitated only by exposure to a circumscribed phobic stimulus.

B. Panic attacks are manifested by discrete periods of apprehension or fear, and at least four of the following symptoms appear during each attack:

1. dyspnea (trouble breathing)
2. palpitations
3. chest pain or discomfort
4. choking or smothering sensations
5. dizziness, vertigo or unsteady feelings
6. feelings of unreality
7. paresthesias (tingling in hands or feet)
8. hot and cold flashes
9. sweating
10. faintness

11. trembling or shaking
12. fear of dying, going crazy, or doing something uncontrolled during an attack

C. Not due to a physical disorder or another mental disorder, such as major depression, somatization disorder or schizophrenia.
D. The disorder is not associated with agoraphobia.

Elaboration

Panic disorder is a chronic condition manifested by attacks of acute anxiety usually occurring in the absence of a fear-provoking situation. It is one of the most common psychiatric disorders, affecting about 5% of the adult population. Women more often have the condition than men.

The above criteria require at least three panic attacks in a three-week period. Actually, panic disorder usually persists for many years, sometimes over a lifetime.

Attacks usually begin suddenly, sometimes in a public place, sometimes at home. There is a sense of foreboding, fear, and apprehension; a sense that one has become seriously ill; a feeling that one's life may be threatened.

Labored breathing, smothering feelings, palpitations of the heart, blurred vision, tremulousness, and weakness usually accompany the apprehension and foreboding. If a patient is examined during a panic attack, signs of distress will be present: rapid heartbeat, sweating, shallow breathing, tremor, hyperactive reflexes, and dilated pupils. An electrocardiogram taken during the episode is usually normal, although the heart rate is often increased (called sinus tachycardia). The chest may be tender from exaggerated breathing.

Panic attacks vary in frequency. Some people experience them daily; others have them only once or twice a year. Between attacks, most feel normal, although some complain of fatigue or headaches.

Heart and respiratory symptoms are the most frequent complaints reported to physicians: "I have heart spells," "I think I'll suffocate," or "I can hardly breathe." The patient usually believes that his illness is medical in nature and the physician often has trouble convincing him otherwise. Emergency rooms in hospitals are frequently visited by people with panic disorder who believe they are having a heart attack.

PANIC DISORDER

A fifty-one-year-old electrical engineer was brought to an emergency room by ambulance complaining of severe left frontal chest pain. His breathing was labored and he complained of numbness and tingling in his lips and fingers. His pulse was 110. He was perspiring heavily and obviously frightened.

The electrocardiogram was normal. His blood chemistry was normal. Nevertheless, the patient was admitted to the intensive care unit. He was observed for twenty-four hours and then discharged free of symptoms.

The patient had first experienced severe chest pain at the age of twenty-one while watching a movie. He had many subsequent episodes of chest pain accompanied by dyspnea (difficulty breathing) and anxiety and had been taken to an emergency room on at least ten previous occasions. Although the examinations were always normal, he had undergone a variety of cardiovascular procedures, including cardiac catheterization. He had never been seen by a psychiatrist and at no time was told that he very likely suffered from a common condition called panic disorder.

Symptoms may become attached to specific situations that the patient then tries to avoid. For example, he may choose aisle seats in theatres, close to exits, so that if an attack occurs, his escape will not be blocked. Or he may avoid social situations in which an attack would be embarrassing. But phobias play a minor role in panic disorder; the attacks usually occur for no apparent reason.

Panic disorder almost always begins in the teens or twenties. Some patients may remember the exact time and circumstance of the first attack. Some remember being awakened at night by their first panic attack. Others remember having their first attack at times of stress, such as making a speech in class. Thus, the disorder may begin suddenly with all the symptoms of a panic attack, but it can also begin insidiously with feelings of tenseness, nervousness, fatigue, or dizziness for years before the first full-fledged anxiety attack is experienced.

The patient's initial medical contact is not always helpful. If he comes to a physician complaining of heart and respiratory symptoms, fearful of heart disease, a physician unacquainted with panic disorder may support the patient's fears by referring him to a specialist and instructing him to avoid exercise.

Sometimes patients complain of problems other than heart and respiratory symptoms. They may have symptoms of "irritable colon," such as abdominal cramping, diarrhea, constipation, nausea, belching, flatus

These usually have Candidiosis, most common cause of panic attacks. Lives in the gut, feeds on sugars, fermenting them to CO_2.

(passing gas), and trouble swallowing, which may prompt them to consult a gastroenterologist.

Panic disorder can be severe but usually the course is fairly mild. Symptoms wax and wane in an irregular pattern that may or may not be associated with events and circumstances interpreted by the patient as stressful. Despite their symptoms, most people with panic disorder live productively without social impairment. Over a 20-year period, more than half of the patients recover or are much improved; about one in five continues to have moderate to severe disability.

Panic disorder rarely leads to hospitalization. It is not associated with suicide. "Going crazy" or doing something uncontrolled during an attack—common apprehensions—are uncommon occurrences. Probably more cardiologists and gastroenterologists see people with panic disorder than psychiatrists. When psychiatrists see them, they are usually suffering from a depression or alcoholism.

Anxiety attacks are occasionally accompanied by hyperventilation, lowering the blood carbon dioxide level and producing symptoms described in Chapter 11. Among some patients there is a disturbing sense that one's body has changed or become distorted—called *depersonalization*. Such a feeling of alien change may extend to the surrounding world—called *derealization*.

Here are ways patients describe depersonalization and derealization, quoted by David Sheehan in his book *The Anxiety Disease*:

> I feel I'm in another world. It's like I know I'm there, but I'm really not. I feel removed from the situation I'm in. I feel like I'm in another dimension—like a hollow or vacuum—outside the situation I'm in. It's like watching the whole thing from a distance . . . like I was looking at everything through the wrong end of a telescope—it seemed to get more distant and smaller. Sometimes the opposite would happen and everything would feel closer and larger . . .
>
> Sometimes I feel disconnected from the ground under my feet, like I'm walking on air or floating. Last week I was driving my car when I felt that the wheels weren't connected to the road. The car seemed as if it was floating along about two feet off the ground. I felt I didn't have control—I couldn't reconnect with the ground properly.

Emily Dickinson, a victim of panic attacks, also experienced depersonalization:

> I felt a cleavage in my mind
> As if my brain had split;
> I tried to match it, seam by seam,
> But could not make them fit.
> The thought behind I strove to join
> Unto the thought before,
> But the sequence ravelled out of reach
> Like balls upon a floor.

Describing derealization, a patient recalls:

> I was lying awake in bed. The curtains were open and the full moon was out. I was a little frightened, so I held onto my husband's arm for security. Then all of a sudden I couldn't feel him or myself any more. I felt as if I was at the other side of the room looking back at myself in bed.

As Sheehan pointed out, such experiences are not unusual to a milder degree among the normal population, particularly during adolescence. "The hand holding the pen doesn't seem like it's yours; the face in the mirror actually seems like it's someone else's." These derealization and depersonalization experiences are usually a little frightening. Patients sometimes think they are losing their mind, but such symptoms are not usually signs of insanity. Most often they are associated with anxiety.

Panic attacks can be a symptom of any psychiatric illness. Anxiety symptoms often appear in depressive disorders, obsessional illness, phobic disorders, hysteria, and alcoholism. In part, the diagnosis is based on chronology. Panic disorder is diagnosed only if there are no other "preemptive" symptoms (Chapter 10) or if anxiety symptoms occurred before other symptoms, such as depression, developed.

Medical illnesses that produce symptoms resembling those of panic disorder include cardiac arrhythmias (especially paraoxysmal atrial tachychardia), angina pectoris, hyperthyroidism, pheochromocytoma, parathyroid disease, hypoglycemia, and mitral valve prolapse. Of these, mitral valve prolapse has received by far the greatest attention in recent years.

Mitral Valve Prolapse

Only in the last 20 years has the significance of mitral valve prolapse become appreciated. Between 5 and 20% of otherwise healthy people, most of them female, are believed to have this condition. Although it rarely causes symptoms, it confuses physicians as well as patients. It appears that at least one-third of people with panic disorder also have mitral valve prolapse, but nobody is clear as to what this means.

Since mitral valve prolapse is a common heart abnormality and closely related to panic disorder, something needs to be said about the condition.

The mitral valve controls blood flow through the chambers on the left side of the heart. The left upper chamber, or atrium, receives blood from the lungs and transfers it down to the heart's main pumping chamber, the left ventricle, which sends the blood to the far reaches of the body. It is the job of the mitral valve to be sure blood flows only in one direction through these chambers.

The valve itself resembles a billowy sleeve. (Its name comes from the word miter, the Bishop's ceremonial headdress it resembles.) As blood passes from the atrium to the ventricle below, the cuff of the sleeve remains open. Then as the lower chamber pumps, it is pulled close to prevent flow back into the atrium. In mitral valve prolapse, one or both flaps of the valve are bulging or enlarged, allowing them to push up into the atrium like an umbrella turned inside out by a gust of wind. This prevents the valve from shutting fully and may allow blood to seep back into the atrium.

Heard through a stethoscope, mitral valve prolapse produces characteristic heart sounds. The normal heart beat sounds like "lub-dub." But as a prolapsed valve fills with blood, it may make a popping noise, or click, between the lub and the dub. If blood seeps back into the atrium, after the click a murmur may be heard that sounds like sawing wood. Another name for the condition is click-murmur syndrome.

To detect the click-murmur, the physician needs to listen to the heart in a variety of different body positions and then confirm the diagnosis with an echocardiogram, an examination of the heart by ultrasound.

The cause of mitral valve prolapse is not known. Recent evidence suggests that it is inherited as a dominant trait (if one parent has it, each child has a 50–50 chance of inheriting it). The syndrome is usually not detected until the teens or later. Many people go through life not knowing they have it.

The symptoms from mitral valve prolapse are identical with those of panic disorder: chest pain (not relieved by nitroglycerin); shortness of

Try treating trigger pts anterior ribs T-5-7 on low fibers of pec major, Rt side causes irregular beat, L side angina symptoms. See Janet Travell Myofascial Pain and Dysfunction

PANIC DISORDER

breath; brief episodes of dizziness, faintness, numbness, and tingling; skipped heart beats; and feelings of anxiety and panic.

As noted, the connection between panic disorder and mitral valve prolapse remains obscure. Is it possible that frequent panic attacks would cause mitral valve prolapse, perhaps by triggering abnormal surges of "stress hormones"? Or do people diagnosed as having panic disorder really only have mitral valve prolapse that mimics the symptoms of panic disorder? Interestingly, infusions of sodium lactate that produce panic attacks in people with a history of panic disorder also produce attacks in those with mitral valve prolapse. Furthermore, drugs that relieve panic attacks for people without mitral valve prolapse also relieve panic attacks for people *with* mitral valve prolapse. These include both antidepressant drugs and propranolol.

In any case, mitral valve prolapse—like panic disorder—is rarely serious and those who have it do not have a shorter life expectancy than those without it. [handwritten: In France a condition called "Spasmophilia" blamed on Mg. deficiency, is similar.]

Depression and Social Class

The most common mistake made in the evaluation of anxiety symptoms is the failure to recognize a depressive disorder. Among other factors, a family history of depression should alert the physician to the possibility that a patient complaining of anxiety symptoms may have a depression. Anxiety symptoms that begin for the first time after the age of forty are commonly part of a depressive syndrome rather than a manifestation of an anxiety disorder.

Patients of lower social class apparently have more severe symptoms and greater social impairment from panic disorder than patients with higher incomes and more education. One explanation is that lower class people are less likely than more affluent people to encounter a physician who takes the time to provide reassurance by explaining the nature of the illness.

Clues to a Cause

The cause of panic disorder is unknown. However, several observations point to a biological predisposition. [handwritten: cause,]

First, persons with panic disorder differ physiologically in several ways from normal individuals. They are more responsive to pain. They tire more rapidly after exercise. The lactic acid in their blood rises to higher levels after exercise.

The latter observation led to experiments where sodium lactate (the salt of lactic acid) was administered to people with panic disorder to see if it would trigger an attack. It did. This is such a reliable finding that sometimes lactate is given to individuals suspected of having panic disorder as a diagnostic test. People without panic disorder do not develop anxiety symptoms after receiving lactate. Little is known about why lactate has this effect in people with panic disorder (see Chapter 5).

Another reason to believe in a biological susceptibility to panic disorder is the fact that it runs in families. Nearly half of people with panic disorder have close relatives with the disorder, compared to a 5% prevalence rate in the general population.

Alcoholism is also often found in the families of people with panic disorder. Since alcoholics commonly have anxiety attacks, particularly when hung over, it is hard to tell which came first—panic disorder or alcoholism. Both typically begin in the teens and twenties.

Studies do *not* find a high prevalence of generalized anxiety disorder or obsessive compulsive disorder in the families of patients with panic disorder. This suggests these disorders are truly separate diagnostic groups with different causes.

Treatment

One of the oldest controversies in psychiatry is whether anxiety neurosis—now called panic disorder—is a variant of clinical depression. Perhaps the strongest argument against panic disorder being a form of "masked" depression is that the mask does such a good job of masking: most people have spontaneous panic attacks off and on for many years and yet never really become depressed in the way, for example, that a person with manic-depressive disease is depressed. Also, depression is associated with a high rate of suicide while panic disorder is not. Furthermore, panic disorder begins early in adult life whereas depression is more common in the middle and late years. Finally, panic attacks can usually be precipitated in people with a history of spontaneous attacks by the intravenous infusion of sodium lactate, and this rarely happens in depressed individuals.

In recent years a new discovery of singular importance has refueled the controversy. It has been found that drugs that relieve depression are strikingly successful in preventing panic attacks, whereas drugs that relieve anxiety have practically no effect on panic attacks. There are two major families of antidepressant drugs—the tricyclic group (such as Elavil) and the MAO-inhibitor group (such as Nardil)—and both prevent panic attacks. (Not all drugs in the two classes have been studied, but, based on present knowledge, it is likely that all have an "anti-panic" effect.)

Now comes a second-generation benzodiazepine—a distant relative of Librium—and *it* seems to prevent panic attacks. Doesn't this contradict the statement that antianxiety drugs do not prevent panic attacks? Ordinarily they do not, but this particular new antianxiety drug—Xanax, or alprazolam—may relieve depression as well as anxiety and also prevent panic attacks.

Here is further support for the surprising effectiveness of antidepressant drugs for panic attacks: tricyclic antidepressants, MAO inhibitors, and the new second-generation benzodiazepine, Xanax, all block panic attacks that are artificially produced by an intravenous infusion of sodium lactate given to people who have the disorder.

A word of warning: just because a drug relieves more than one condition does *not* mean that these conditions are the same. Drugs relieve congestive heart failure by acting on the kidney, but this does not mean the kidney has caused the congestive heart failure. There are numerous examples in medicine where drugs relieve a condition without acting directly on the cause of the condition.

Thus, the unexpected finding that antidepressant drugs are the closest thing we have to a definitive treatment for panic disorder does not mean that panic disorder is a form of depression.

Although it is generally accepted that the best treatment for panic disorder available today is one of the antidepressant drugs, there is much controversy about which drugs are best and the most effective dosage. Ordinarily, when one investigator finds that a drug works and another reports that it does not work, the former accuses the latter of using too low a dosage. Xanax, for example, apparently requires much higher doses to prevent panic attacks than those recommended by the manufacturer.

Here is where a psychiatrist may be needed, not to administer psychotherapy but to bring to bear his or her expertise in psychoactive drugs.

Everything else being equal, psychiatrists generally know more about drugs that affect emotions than nonpsychiatrists do. Giving the right drug to the right person in the right dosage is still perhaps more of an art than science. Yet knowing the chemical properties of the drugs—their effects, side effects and effectiveness—is a prerequisite for effective use of these drugs.

As with generalized anxiety disorders, supportive psychotherapy, described in Chapter 11, is of great importance in treating panic disorder. The value of education and reassurance cannot be overestimated, but neither can the value of one of the antidepressant medications in preventing attacks or reducing their severity.

The beta-adrenergic blocking agent propranolol, a drug used for irregular heart beats, has also been used in treating panic attacks. Although its efficacy has not been established definitely, there is little question that it blocks some of the physiologic symptoms.

13

Phobia: An Introduction

> Love casts out fear; but conversely fear casts out love. And not only love. Fear also casts out intelligence, casts out goodness, casts out all thought of beauty and truth . . . in the end fear casts out even a man's humanity.
>
> Aldous Huxley
> *Ape and Essence*

A phobia is a persistent, excessive, unreasonable fear of a specific object, activity, or situation that results in a compelling desire to avoid the dreaded object, activity, or situation. The fear is recognized by the individual as excessive and unreasonable. The avoidance behavior involves some degree of disability.

This definition has four essential features. Phobias are *persistent*. Fear of remounting a horse after falling off one is not a phobia unless the fear persists and leads to the second essential feature: *avoidance* of the feared object, activity, or situation (in this case, horses). Fears that do not lead to avoidance usually go away, assuming they are unrealistic. This fortunate fact of life, indeed, is the basis for most treatments of phobia.

The third element: to be phobic, the fear must not only seem *unreasonable* to others but be viewed as unreasonable by the victim. When a schizophrenic has an irrational fear, he usually does not recognize its irrationality, in which case the fear is not a phobia but a delusion (defined by psychiatrists as a fixed, false idea).

The fourth element is decisive in separating "normal" fears from phobias. How *disabling* is the fear? Irrational avoidance of objects, activities, or situations is common. If the effect on life adjustment is insignificant, it little matters. Many people fear and avoid harmless insects and spiders, but it does not affect their lives. However, when fear and avoidance significantly interfere with a person's mental functioning or social adjustment, it is excessive and therefore phobic.

The four elements of a phobia are illustrated in this account by a phobic psychiatrist:

> I was pampering my neurosis by taking the train to a meeting in Philadelphia. It was a nasty day out, the fog so thick you could see only a few feet ahead of your face, and the train, which had been late in leaving New York, was making up time by hurtling at a great rate across the flat land of New Jersey. As I sat there comfortably enjoying the ride, I happened to glance at the headlines of a late edition which one of the passengers who had boarded in New York was reading. TRAINS CRASH IN FOG, ran the banner headlines, 10 DEAD, MANY INJURED. I reflected on our speed, the dense fog outside, and had a mild, transitory moment of concern that the fog might claim us victim, too, and then relaxed as I picked up the novel I had been reading. Some minutes later the thought suddenly entered my mind that had I not "chickened out" about flying, I might at that moment be overhead in a plane. At the mere image of sitting up there strapped in by a seat belt, my hands began to sweat, my heart to beat perceptibly faster, and I felt a kind of nervous uneasiness in my gut. The sensation lasted until I forced myself back to my book and forgot about the imagery.
>
> I must say I found this experience a vivid lesson in the nature of phobias. Here I had reacted with hardly a flicker of concern to an admittedly small but real danger of accident, as evidenced by the fog-caused train crash an hour or two earlier. At the same time I had responded to a purely imaginary situation with an unpleasant start of nervousness, experienced both as physical symptoms, and as an inner sense of indescribable dread so characteristic of anxiety. The unreasonableness of the latter was highlighted for me by its contrast with the absence of concern about the speeding train, which if I had worried about it, would have been an apprehension grounded on real, external circumstances.

Because a phobia involves a cluster of elements, it is called in medical parlance a disorder. Actually, there are three major phobic disorders: *simple phobia,* *social phobia,* and *agoraphobia.* Each has the essential four features of a phobia and also what doctors call a characteristic natural history. This means the disorder has a specific age of onset, course, and outcome. In this sense, a phobic disorder is like measles: there is information about when it begins, what it does, and how it ends. Is a disorder the same as a disease? If one defines disease as a distressing condition that leads people to consult physicians, it meets the definition of disease.

Prevalence

How common is phobia? There are two ways to find out. One is to study a random sample of the population. The other is to study people who go to doctors. The second method is unsatisfactory. People see doctors for other reasons than their illness. They see doctors because doctors are available, because they have the money to see doctors, because seeing doctors is fashionable, or for other reasons.

A recent survey published by the National Institute of Mental Health has found that roughly one adult in 20 suffers from the most serious variety of phobia, agoraphobia, and that one in nine adults harbors some kind of phobia, making it this country's most common mental health problem. The presence of agoraphobia and social phobia in two large series of psychiatric patients with predominant mood disorders has been reported to range from 50 to 65%, and psychiatrists see large numbers of patients with "predominant mood disorders." One of these studies had a control group consisting of patients from a fracture clinic where the prevalence of agoraphobia and social phobia was found to be 30% and 16%, respectively. However, these figures may have been spuriously elevated due to the frequency of alcohol abusers in a fracture clinic population. One study found that alcohol abusers have a particularly high rate of phobia, one-third suffering from disabling phobias and another third from milder phobias.

Here are two more reasons for the discrepancy in prevalence data about phobias: (1) some physicians may include what others call "normal fears" in their statistics and (2) people tend to be secretive about phobias. Many people are embarrassed about their phobias and often keep them hidden from even their closest friends. This habit of secrecy may persist even when they see doctors and see them for psychiatric reasons. It is not unusual in psychiatric practice to see a patient for a long period and then have him describe, almost in passing, a phobia that has plagued him for years. In routine questioning of patients, phobias often are not asked about. Many nonpsychiatrists seem unaware even of their existence, or at least of their possible medical significance.

Historical Background

Phobos was a Greek god called upon to frighten one's enemies. His likeness was painted on masks and shields for this purpose. Phobos, or phobia, came to mean fear or panic.

"Phobia" first appeared in medical terminology in Rome 2,000 years ago, when "hydrophobia" (fear of water) was used to describe a symptom of rabies. Though the term was not used in a psychiatric sense until the nineteenth century, phobic fears and behavior were described in medical literature long before that. Hippocrates described at least two phobic persons. One was "beset by terror" whenever he heard a flute, while the other could not go beside "even the shallowest ditch" and yet could walk in the ditch itself.

Robert Burton in *Anatomy of Melancholy* distinguishes "morbid fears" from "normal fears." Demosthenes' stage fright was normal, while Caesar's fear of sitting in the dark was morbid. Burton believed normal fears could be overcome by willpower, but not morbid fears.

The term phobia appeared increasingly in descriptions of morbid fears during the nineteenth century, beginning with "syphiliphobia," defined in a medical dictionary published in 1848 as "a morbid dread of syphilis giving rise to fancied symptoms of the disease." Numerous theories were advanced to explain phobias, including poor upbringing.

In 1871, a German neurologist named Karl Westphal described three men who feared public places and labeled the condition agoraphobia, "agora" coming from the Greek word for place of assembly or marketplace. Historians give Westphal credit for first describing phobia in terms of a *disorder* rather than an isolated symptom. Westphal even prescribed a treatment for the condition, suggesting that alcohol, a companion, or the use of a walking cane in public places would be helpful.

Later investigators compiled long lists of phobias, naming each in resounding Greek or Latin terms after the object or situation feared. Thus, as the contemporary psychiatrist John Nemiah points out, "The patient who was spared the pangs of taphaphobia (fear of being buried alive) or ailurophobia (fear of cats) might yet fall prey to belonophobia (fear of needles), siderodromophobia (fear of railways), or triskaidekaphobia (fear of thirteen at table), and pantaphobia was the diagnostic fate of that unfortunate soul who suffered from them all."

Since the late nineteenth century there has been a continuing controversy over the relationship of phobias to other psychiatric disorders.

PHOBIA: AN INTRODUCTION

The famous German psychiatrist, Emil Kraepelin, spoke of phobias and obsessions as though synonymous. The even more famous Sigmund Freud separated phobic neurosis from obsessive compulsive neurosis and anxiety neurosis—a separation that still prevails in most psychiatric textbooks. Some psychiatrists regard all phobias as a manifestation of manic-depressive disease. The majority, however, agree that phobias may occur in various psychiatric conditions, but may also be the primary manifestation of specific phobic disorders.

DSM-III—the new classification of mental disorders—separates phobias into three disorders: *simple phobia, social phobia, and agoraphobia*. The next three chapters deal with each separately, followed by a chapter on *school phobia* and a chapter discussing the treatment of phobia in general.

14

Simple Phobias

There were 117 psychoanalysts on the Pan Am flight to Vienna and I'd been treated by at least six of them. . . . God knows it was a tribute either to the shrinks' ineptitude or my own glorious unanalyzability that I was now, if anything, more scared of flying than when I began my analytic adventures some thirteen years earlier.

My husband grabbed my hand therapeutically at the moment of takeoff.

"Christ—it's like ice," he said. . . . My fingers (and toes) turn to ice, my stomach leaps upward into my rib cage, the temperature in the tip of my nose drops to the same level as the temperature in my fingers . . . and for one screaming minute my heart and the engines correspond. . . . I happen to be convinced that only my own concentration . . . keeps this bird aloft. . . . I congratulate myself on every successful takeoff, but not too enthusiastically because it's also part of my personal religion that the minute you grow overconfident and really *relax* about the flight, the plane crashes instantly . . .

Erica Jong
Fear of Flying

A simple phobia is simply that: an isolated fear of a single object or situation, leading to avoidance of the object or situation. The fear is irrational and excessive but not always disabling because the object or situation can sometimes be easily avoided (for example, snakes, if you are a city dweller). The DSM-III criteria for simple phobia follows.

Diagnostic Criteria

A. A persistent, irrational fear of, and compelling desire to avoid, an object or a situation other than being alone, or in public places away

from home (agoraphobia), or of humiliation or embarrassment in certain social situations (social phobia). Phobic objects are often animals, and phobic situations frequently involve heights or closed spaces.
  B. Significant distress from the disturbance and recognition by the individual that his or her fear is excessive or unreasonable.
  C. Not due to another mental disorder, such as schizophrenia or obsessive compulsive disorder.

Elaboration

Impairment may be considerable if the phobic object is common and cannot be avoided, such as a fear of elevators by someone who must use elevators at work. Fear of flying—the bane of Erica Jong and the phobic psychiatrist in the preceding chapter—is, to say the least, inconvenient in the jet age, and devastating for national politicians and traveling salesmen.

Simple phobia is not usually associated with other psychiatric symptoms or other psychiatric disorders, such as depression. The phobic person is no more (or less) anxious than anyone else until exposed to the phobic object or situation. Then he becomes overwhelmingly uncomfortable and fearful, sometimes having symptoms associated with a panic attack (palpitations, sweating, dizziness, difficulty breathing). The phobic person becomes afraid just thinking about the possibility that he might be confronted with the phobic stimulus. This is called anticipatory anxiety; it leads people to avoid situations in which the phobic stimulus *might* be present.

Simple phobia rarely leads to medical consultation. Of the 11 million Americans who have significant phobias (according to a recent study by the National Institute of Mental Health), about two-thirds have simple phobias, but there is reason to believe simple phobias are more common than this.

For example, when a new and apparently effective treatment for simple phobias was introduced by Joseph Wolpe in 1958, people with simple phobias started showing up in doctors' offices. When the Maudsley Hospital in London ran newspaper advertisements announcing a phobia clinic, patients came in droves. Some did not have a phobia or much of a phobia; they were lonely and wanted to talk to someone; or they had a depression or something else wrong and did not know who to see.

But many had true phobias they had never mentioned to anyone, much less to a doctor.

If phobias were as common as *names* for phobias, they would be more common than the common cold. The reason for so many names is this: people can become phobic about almost any object or situation. Classifying phobias by the feared object or situation—a common practice in the nineteenth century—can be performed by anyone with a little Greek or Latin, as shown in the table (compiled by psychiatrist Isaac Marks to give a taste of what is possible).

The more common simple phobias include fear of animals, heights, closed spaces, doctors and dentists, wind, storms, lightning, loud noises, driving a car, flying in planes, traveling by subway, injections, and the sight of blood.

Fear of crowds is also common, but most often seen in agoraphobia, discussed later. A fear of public speaking is very common but comes under the category of social phobia, also discussed later.

There are endless uncommon phobias, including fear of running water, of swallowing solid food, and of going to the hairdresser. There is even the case of the tennis player who wore gloves because he had a phobia about fuzz, and tennis balls are fuzzy. To repeat: phobias can develop toward any object or situation. However, some phobias are more common than others and apparently occur in every society. The fears in part may be innate.

Children seem born with a fear of strangers and a fear of being looked at. This may explain why so many people are afraid of public speaking; the old inherited fears never quite go away and are often reinforced by experiences in later life. Fears of darkness, bats, and other undomesticated creatures may be a carryover of childhood fears of monsters lurking in the dark and may also be based on instinctive fear of objects and situations combining familiar and unfamiliar elements (Chapter 4). Dead people look *almost* like live people but not quite, and are frightening to children.

Even when phobias in adults have no apparent connection with instinctive fears of childhood, many at least seem understandable (more understandable anyway than the man who won't touch tennis balls). Injections hurt a little; dentists can hurt a lot. Nevertheless, when people who eagerly want to visit a foreign country stay home because they cannot stand the thought of having a smallpox shot, this is a phobia. When

SIMPLE PHOBIAS

Formal names given to some phobias

Acrophobia	height (Gr. *acro*, heights or summits)
Agora	open spaces (Gr. *agora*, marketplace, the place of assembly)
Ailuro	cats (Gr. *ailuros*, cat)
Arachno	spiders (Gr. *arachin*, spider)
Antho	flowers (Gr. *anthos*, flower)
Anthropo	people (Gr. *anthropos*, man generically)
Aqua	water (Lat. *aqua*, water)
Astra	lightning (Gr. *asterope*, lightning)
Bronto	thunder { (Gr. *bronte*, thunder)
Keraunos	(Gr. *keraunos*, thunderbolt)
Claustro	closed spaces (Lat. *claustrum*, bar, bolt, or lock)
Cyno	dogs (Gr. *cynas*, dog)
Demento	insanity (Lat. *demens*, mad)
Equino	horses (Lat. *equus*, horse)
Herpeto	lizards, reptiles (Gr. *herpetos*, creeping or crawling things)
Mikro	germs (Gr. *mikros*, small)
Muro	mice (Lat. *murmus*, mouse)
Myso	dirt, germs, contamination (Gr. *mysos*, uncleanliness, abomination)
Numero	number (Lat. *numero*, number)
Nycto	darkness (Gr. *nyx*, night)
Ophidio	snakes (Gr. *ophis*, snake)
Pyro	fire (Gr. *pyr*, fire)
Thanato	death (Gr. *thanatos*, death)
Tricho	hair (Gr. *tricho*, hair)
Xeno	stranger (Gr. *xeros*, stranger)
Zoo	animal (Gr. *zoos*, animal)

people walk around with a toothache because dentists terrify them, this is a phobia.

Animal phobias are the most common type of simple phobia. Two other common simple phobias are fear of heights and fear of illness. All three categories probably involve instinctual elements and all three are somewhat understandable in that animals can be dangerous, people do fall off high places and illnesses are no fun. When these fears become exaggerated and unreasonable, however, they are distressing and even disabling. Each category deserves some additional comment.

Animal Phobias

These phobias have been studied more than any other. They almost always begin in childhood, often before the age of seven. They usually subside before puberty, but many adults continue to have animal phobias well into adulthood and often to the grave.

Most people are phobic of only one species. This may be cats, dogs, horses, other domesticated animals, birds, snakes, frogs, fish, and wild animals like bats. Many people are actively afraid of, or at least squeamish about, worms, mice, and spiders, but the fearfulness is easily controlled and they do not actively avoid even the *possibility* of encountering one of these animals. Sometimes the phobia is directed toward one feature of the animal, such as feathers.

Impairment is associated with animal phobias when the animals cannot be avoided. For example, some city people have phobias toward pigeons and stay off the streets to avoid them. With respect to frogs, fish, and snakes, these can usually be avoided, sometimes at the expense of foregoing possible pleasure and even possible employment. Spiders? Avoidance is not so easy. The distress may be striking.

> A woman with a fear of spiders screamed when she found a spider at home, ran away to find a neighbor to remove it, trembled in fear and had to keep the neighbor at her side for two hours before she could remain alone at home again; another patient with a spider phobia found herself on top of the refrigerator in the kitchen with no recollection of getting there—the fear engendered by sight of a spider had induced a brief period of amnesia. (Reported by Isaac Marks.)

Animal phobias do not involve a fear of contamination by the animal from fleas, dirt, parasites, etc. People with exaggerated fears of contamination (from animals or other sources) almost always have obsessional neurosis (Chapter 19).

Unlike most psychiatric disorders, animal phobias do not run in families. Little girls with cat phobias as a rule do not have parents with cat phobias. This has two implications: (1) animal phobias are not directly influenced by heredity, and (2) children do not learn them from their parents. So why do they occur?

SIMPLE PHOBIAS

Sometimes animal phobias seem related to a specific event. Father takes the children out to watch him drown the kittens, and at least one of them develops an intense desire to avoid kittens. A child gets bitten by a ferocious dog and thereafter avoids dogs, ferocious or otherwise.

However, most phobias seem unrelated to stressful events. Even Sigmund Freud, one of the world's greatest explainers, had trouble explaining animal phobias: He wrote in 1913:

> The child suddenly begins to fear a certain animal species and to protect itself against seeing or touching any individual of this species. There results the clinical picture of an animal phobia, which is one of the most frequent among the psychoneurotic diseases of this age and perhaps the earliest form of such an ailment. The phobia is as a rule expressed towards animals for which the child has until then shown the liveliest interest, and has nothing to do with the individual animal. In cities, the choice of animals which can become the object of phobia is not great. They are horses, dogs, cats, more seldom birds, and strikingly often very small animals like bugs and butterflies. Sometimes animals which are known to the child only from picture books and fairy stories become objects of the senseless and inordinate anxiety which is manifest in these phobias. It is seldom possible to learn the manner in which such an unusual choice of anxiety has been brought about.

Of all phobias, people with animal phobia probably respond best to types of behavior therapy described in Chapter 18.

Fear of Heights

Fear of heights is really a fear of falling. It may be totally unrealistic. The person may be close enough to the ground for a fall not to hurt him, or be sufficiently protected by the environment for a fall to be impossible. No amount of reasonableness, however, curtails the fear. It is rooted in a basic instinct, the fear of receding edges discussed in Chapter 4 and shared by young goats and small children alike. Most children and goats lose the fear, but not all, and a very disabling fear it can be.

Some victims will not walk down a flight of stairs if they see the open stairwell. Some will not look out a window from the second floor or above, particularly if the window goes from floor to ceiling. Others will not cross a bridge on foot, although they may cross it by car without worrying.

Just as a fear of heights is really a fear of falling, a fear of falling is really a fear of loss of support. More exactly, it is fear of loss of *visual* support. Hence the person looking out a window experiences no fear if the window is waist-high. A car offers the same protection.

Behind the fear of losing visual support seems to be an even deeper fear: the fear of being *drawn* over the edge of the height. The high-up window, the car's doors, provide a sense of protection that no amount of reasoning-with-oneself can give.

Surprisingly, a fear of heights is rather frequently reported by airline pilots. As a consultant to the US Pilots Union, I have had a number of pilots tell me they panic looking over the top of buildings—but fly at 40,000 feet with nonchalance. Again, it is not the height that counts, but the "visual space." There is not much visual space in a plane, even in the cockpit.

Illness Phobias

We all have a fear of illness, though stoics may not admit it or show it. The lump, the cough, the spot on the skin: probably nothing, but who can be sure? Some people rush off to doctors at the first hint of trouble; most do not. They just worry for a while and hope it goes away. It usually does.

But a few continue worrying—even after the lump, the cough, the spot have disappeared. The fear becomes persistent, excessive, irrational: a phobia.

Illness phobias change with the times. For centuries, phobia of venereal disease was common. Generations of men inspected their genitalia for minute changes, and, finding none, still worried. They worried even if there had been no sexual exposure; virginal bachelors seemed particularly susceptible. There was always an explanation: toilet seats have been blamed as long as there have been toilets (and have always been as blameless). Doctors, finding nothing, shrugged or gave yesterday's equivalent of a vitamin B shot, which did not help phobic patients then any more than it does now.

Coughs are common, and so was tuberculosis phobia until a cure came along and people stopped talking about TB. Today, cancer phobia leads the list, with heart disease a close second. Growlings, gurglings, rumblings: all the normal cacophony of the intestinal tract invoke terror in the cancer phobic.

Illness phobics go to doctor after doctor, have examination after examination—and, no matter how reassuring the former and negative the latter, are still not convinced. Excessive reassurance is particularly viewed with alarm.

Some illness phobics do the opposite: avoid doctors at all costs. They refuse to take insurance examinations because of what might be found. Suspecting a lump in a breast, they never touch the breast, even when bathing. If the disease they fear is ultimately found to exist, the fear often goes away. "Fear is more pain than the pain it fears," Sir Philip Sidney said.

Ironically, fear of illness may produce, if not illness, symptoms of illness. Fear both tightens and loosens the sphincter muscles, causing constipation, diarrhea, or both in rotation. Fear makes the heart pound. Fear tightens scalp muscles, the real cause of most headaches self-interpreted as incontrovertible evidence of brain tumor.

Fear, some believe, actually causes some illnesses such as asthma and colitis, but the evidence for this is weak. Fear undoubtedly worsens the symptoms or even precipitates attacks (this is not the same as causing an illness).

People with illness phobia often know friends and relatives who have had the feared illness, perhaps explaining the choice of the particular imagined illness but hardly its intensity and intractability.

Illness phobia needs to be distinguished from hypochondriasis. Hypochondriacs have many imagined symptoms and illnesses. They usually take their doctor's word that it is imaginary or "psychological" or whatever euphemism used, only to return on another day with another imaginary illness.

Illness phobia fixes on one illness, and no amount of reassurance unfixes it.

More men than women have illness phobias; more women are hypochondriacs (or at least see doctors repeatedly for illnesses that they don't have).

Illness phobia also must be distinguished from delusions. Both phobias and delusions are fixed, false ideas. However, the delusional patient does not realize the idea is false. The phobic patient will grant that it is foolish, but be terrified anyway. Delusions occur in schizophrenia, manic-depressives, and brain diseases. Simple phobias are isolated symptoms in otherwise healthy individuals.

How common are illness phobias? Of simple phobias seen by psychiatrists (hardly a representative sample of the general population), 15 to 30% are illness phobias.

SIMPLE PHOBIAS usually begin early in life. Animal phobias almost always begin before puberty and usually in early childhood. Other simple phobias may begin somewhat later but most begin at least by the mid or late twenties. An exception is illness phobia, which often begins in midlife.

Sometimes phobias develop so gradually that the victims have trouble remembering precisely when they began. Other victims recall vividly the onset of a phobia, particularly if it was accompanied by a panic attack. Many years later, they can identify the hour, day, precise location where it happened.

Simple phobias occur in men and women about equally, with animal phobias occuring more in women and illness phobias more in men. More women are *treated* for phobias than men, but this may be because more women see doctors, regardless of the problem.

Once a phobia has lasted for a year, it tends to become chronic, meaning the fear may persist for many years unless two things happen: (1) the patient's life circumstances change so that he or she *must* confront the feared object or situation repeatedly, in which case the phobia has no "choice" other than retreat in the face of necessity; (2) the patient receives treatment.

The first happens commonly. Trains are discontinued and the salesman with phobia about flying *must* travel by plane to keep his job. The recession worsens and the only job the young man can find is selling encyclopaedias, which he does in spite of his phobia about strangers. In other words, when the choice is avoidance or survival, the healthy person with an isolated phobia choses survival—and after a time the fear goes away (although not always completely).

Confrontation of the feared object or situation is the foundation of the behavior treatment of simple phobia (Chapter 18). The effectiveness of harsh reality (discontinuation of trains, recessions) in "curing" phobias is undeniable.

In three studies, phobic individuals have been studied over a period

of time to see what happens to them and their phobias. Here are the conclusions:
1. Phobias that begin before the age of 20 tend to improve and eventually disappear. Five years after the phobia begins, about half of the young victims are symptom-free and almost all are improved (less bothered). Phobias beginning after adolescence continue for longer periods, with about half of patients improved after five years but only about 5% symptom-free. In the older group, the phobia gradually becomes more severe in about one-third of patients. About 20% are impaired in their work or social functioning because of the phobia. Gender has little influence on the outcome of simple phobia: women do as well (or poorly) as men, with age of onset being the most important prognostic indicator.
2. Simple phobias as a group have a more favorable prognosis than the other phobic disorders. One reason is that simple phobias almost always involve a single isolated phobia whereas social phobias more often involve two or more phobias and agoraphobia invariably involves multiple phobias.

An exception is illness phobia. It typically occurs over the age of 40, when "normal" fear of cancer and heart attacks is endemic (at least in the U.S.). The late age of onset is untypical of simple phobias and carries a worse prognosis. The phobia, if anything, gets worse rather than better, and illness phobics are notoriously hard to treat. The illness phobic cannot avoid the feared object because there *is* no feared object. There is no possibility for the phobia to be cured by direct confrontation.
3. Patients with simple phobias come from reasonably stable families and have reasonably normal childhoods and marriages.

Simple phobia, in conclusion, is a rather benign disorder, particularly if it involves only a single phobia and begins in childhood or adolescence. Although the phobia may persist for many years, it rarely gets worse and often gets better and is not associated with more serious psychiatric disorders. The phobic anxiety is usually quite manageable and impairment slight, either because the dreaded object or situation can mostly be avoided or because, if unavoidable, the object or situation stops being dreaded.

15

Social Phobias

> Fear of two staring eyes is ubiquitous throughout the animal kingdom.
>
> Isaac Marks

A social phobia is basically a fear of being looked at. The fear may be partly innate (see Chapter 4). Monkeys and other animals dislike stares. But when stares, real or imagined, produce extreme discomfort in *particular situations*, the result is a social phobia: "social" because scrutiny by other prople is always in some way involved. The DSM-III criteria for social phobia follows.

Diagnostic Criteria

A. A persistent irrational fear of, and compelling desire to avoid, a situation in which the individual is exposed to possible scrutiny by others and fears that he or she may act in a way that will be humiliating or embarrassing.

B. Significant distress because of the disturbance, and recognition by the individual that his or her fear is excessive or unreasonable.

C. Not due to another mental disorder, such as major depression or avoidant personality disorder.

Elaboration

Behind the fear of scrutiny is a fear of being embarrassed, ridiculed, or making a fool of oneself. And behind this fear is a *performance fear*—the fear of being unable to perform or fear of losing control in some way when others are looking on.

Performance fear often produces what is most feared: poor performance. The panic-stricken public speaker cannot utter a word; the person afraid to eat in public gags at the sight of food.

In some instances the social phobia is the fear of being observed to have a social phobia: a telltale blush or tremor.

Avoidance of the feared situation—the *sine qua non* of any phobia—usually results in inconvenience and distress and, rarely, incapacitation.

Here are some social phobias, starting with the most common.

Fear of Public Speaking

A little stage fright is considered useful. Many professional performers—lecturers, singers, actors—say they have to be a little nervous to be at their best. But extreme stage fright is another matter. It leads to avoidance of stages. Many people can go all their lives without the risk of having to give a talk in public, but often at a cost. The talented singer never sings, the born teacher never teaches. The salesman never becomes a sales supervisor because he would be expected to speak before groups. In its most extreme, fear of public speaking may lead parents to avoid PTA meetings or Sunday school. They might be called upon to say something.

Fear of Using Public Toilets

Performing bodily functions (as the euphemism goes) is not as simple as it might seem. It calls for some intricately timed and synchronized tightening of some muscles and loosening of others. Moreover, performing appropriately—i.e., in the right location at the right time—is highly valued in sanitation-conscious middle-class Western civilization (more so than in older, perhaps just-as-civilized parts of the world).

How common are toilet phobias? Nobody knows, but any army sergeant in a training camp will tell you that constipation is common during the first few days of an inductee's service. Whether lack of toilet privacy is the reason is not known.

Perhaps a more common toilet fear involves apprehension about urinating in public. Many men become anxious standing at a urinal with a line of men waiting behind them. Their sphincters tighten and they become embarrassed by what seems the long delay. Sometimes they depart the urinal unrelieved to avoid attracting more attention.

One reason to think this phobia is fairly common is a scene often observed in large male lavatories at sporting events. Long lines form behind the closed booths despite the fact that urination is the only apparent goal and there are open spaces along the row of urinals. One interpretation is that a sizable number of men have at least a mild social phobia about the act of urination.

What about women? Again, no one knows the frequency of elimination phobias in women, but gynecologists mention women who bring a urine specimen with them for an office visit in case one is needed, fearing they cannot "go on demand" in the doctor's toilet.

Fear of Eating in Public

Here is a case history:

> An attractive 30-year-old wife of a business executive refused to entertain clients and associates of her husband with a dinner party or attend dinner parties in restaurants or someone else's home. Her husband made excuses for her but felt his career was being damaged by their lack of social life. At first her explanation for refusing to join in dinner parties was that it was a waste of time or too much trouble. Later she confessed, tearfully and with much embarrassment, that she was afraid of being *unable* to eat if strangers were watching. Questioned by her concerned husband, she said she was afraid that once food was in her mouth she would be unable to swallow it and then have the embarrassing situation of not knowing where to dispose of the food. She was also afraid that she would gag on the food and possibly vomit.
>
> The phobia began in her late teens when she found the sound of other people eating offensive. Then she became afraid that her own chewing and swallowing would offend others and made a great effort to chew and swallow quietly. This developed into a fear of swallowing any food at all and later a concern that if she did swallow she would vomit, for her the ultimate in shameful acts. She could eat in front of her husband with no difficulty until she told him about the phobia and then became concerned that *he* was watching her eat and expecting her to gag or vomit. From then on she ate alone in the kitchen. The marriage survived, but barely.

Fear of eating in public does not involve loss of appetite. Weight loss may occur but not because the person wants to lose weight or has an obsession with being thin, as happens in anorexia nervosa. In a sense, patients with anorexia nervosa have a severe food phobia, but anorexia nervosa is accompanied by physiological changes such as cessation of menses and bizarre behavior such as hoarding food and self-induced vomiting.

Sometimes fear of eating and drinking in front of others is related to concern that one's hands may tremble while bringing the fork or cup to mouth. The fear of dropping food or spilling coffee in truth is a performance fear, a desire to avoid calling attention to a perceived weakness.

Eating phobias in one way resemble toilet phobias. Both involve semi-automatic functions involving reflexes and muscles of the alimentary canal. These reflexes and muscles are sensitive to the effects of strong emotion such as fear and anger. Fear and anger may cause gagging and tighten the anal sphincter muscle. When a social phobia involves fear of performance based on these delicate mechanisms, the fear itself increases the chance of nonperformance.

Fear of Sexual Performance

Sexual malfunctioning is not usually considered a form of phobia, but, in fact, certain forms of sexual malfunctioning fit the concept of social phobia. The person is anxious about performing because someone is observing his or her performance; the performance anxiety itself increases the chance of nonperformance.

Premature ejaculation, impotence in the male, and frigidity in the female, are the most common forms of social-sexual phobia. Numerous books have been devoted to the subject and it will not be explored here other than to say that the treatment for a social-sexual phobia resembles the treatment for other types of phobia. By various means, performance anxiety is gradually attenuated so that performance can occur.

This type of phobia also involves the essential feature of any phobia, namely, avoidance of the feared situation. Fear of sexual performance probably is one of the world's most effective forms of birth control.

Fear of Being Watched at Work

How many times have you heard: "Don't look over my shoulder while I'm working"? It happens often enough to suggest that this is a common social phobia.

It is truly a phobia only if it leads to some incapacity. If the secretary cannot type or the computer operator cannot punch keys if someone is watching (or may be watching), this reduces output and threatens job security. The phobia can apply to any mechanical operation: taking shorthand, writing on a blackboard, sewing, knitting, or even buttoning a coat if somebody is watching. Teachers need to write on blackboards and seamstresses need to sew. Social phobias are always an inconvenience but sometimes the consequences are serious indeed.

Some people become phobic about writing or handling money in front of others. For example, a person might wait outside a bank until it is empty so he can deposit money. Such fears are often related to a fear of trembling.

"Writer's block" may be phobic, or may reflect some other problem (depression, lack of motivation). The graduate student who has completed his work toward a degree and cannot face writing his thesis may have a social phobia; what he cannot face is scrutiny.

Fear of Crowds

The person with a social phobia of crowds must be distinguished from a person with agoraphobia (Chapter 16). Social phobia of crowds in agoraphobia involves a fear of being enclosed or suffocated. The latter is far more serious than a social phobia, where the person can usually engage in almost any activity as long as nobody is watching. There may be inconvenience. For example, crowd phobia may restrict a person's activities so that he only goes shopping at hours when few people are around and avoids sporting events. It may involve a fear of eye contact; he may stand up in the train to avoid eye contact with sitting passengers.

However, a social phobia is usually not crippling, assuming there is only a single phobia. Agoraphobia involves multiple phobias and is crippling indeed. Hippocrates had a patient with a crowd phobia:

> Through bashfulness, suspicion, and timorousness, he will not be seen abroad; loves darkness and cannot endure the light; his hat still in his eyes, he will neither see, nor be seen. He dare not come in company for fear he should be misused, disgraced, overshoot himself in gesture or speeches, or be sick; he thinks every man observes him . . .

Of course, this may be paranoid schizophrenia. If the patient knows his fear is absurd, it is a phobia; if he considers it justified, it is a delusion and a symptom of schizophrenia or some other psychosis.

Social phobias are often accompanied by a fear of fainting, stumbling, dropping things, or otherwise attracting attention. Phobic people rarely do these things. The only phobia really associated with fainting is a phobia about blood. For some reason, the sight of blood sometimes slows the heart, resulting in a faint (many doctors have seen it happen).

Fear of Being Touched

Just as some people avoid being watched in certain situations, others avoid being touched.

A 23-year-old jazz pianist could tolerate being touched by his girlfriend but by no one else. He played the saxophone as well as piano, but limited his playing to the piano where he would be less likely to be touched. He never played with a traveling group because touching was unavoidable with others in a car or van. He stayed out of elevators, not because of fear of closed places but because of fear of being touched.

He could not explain why touching bothered him. He was not concerned about dirt or contamination from others. He did not associate the fear with sexual contact. He just hated being touched and had felt that way since 15 or 16 when the fear started for no apparent reason.

Once he declined to ride to a sauna to get a massage when he learned there would be other passengers in the car. He looked forward to the massage but the thought of brushing someone's shoulder on the way was intolerable.

Reciprocal inhibition—the psychological principle that holds that one cannot experience two opposing emotions simultaneously—may explain why the pleasure from his girlfriend's touch overrode the incompatible fear of being touched (see Chapter 5).

Miscellaneous Social Phobias

A fear of practicing musical instruments because the neighbors will hear mistakes; a fear of swimming or undressing in front of others because of shame of one's appearance; a fear of driving automobiles; a fear of criticism from superiors (the real reason why some people refuse to work for anyone but themselves).

Most social phobias first occur in adolescence or the early twenties. They rarely begin before puberty or after 30. Social phobias appear to affect both sexes about equally.

Most social phobias develop over several months, stabilizing over a period of years with a gradual diminution of severity in middle life. Most begin without any apparent precipitating event, such as loss of a family member or emotional trauma. Social phobias do not run in families, so there is no evidence for a hereditary factor. In truth, social phobias are a mystery. They have been little studied and their cause is unknown.

16

Agoraphobia

> When a trout rising to a fly gets hooked on a line and finds himself unable to swim about freely, he begins a fight which results in struggles and splashes and sometimes an escape. Often, of course, the situation is too tough for him.
> In the same way the human being struggles with his environment and with the hooks that catch him. Sometimes he masters his difficulties; sometimes they are too much for him. His struggles are all that the world sees and it usually misunderstands them. It is hard for a free fish to understand what is happening to a hooked one.
>
> Karl Menninger
> *The Human Mind*

In 1871 a German neurologist named Karl Westphal described three male patients who shared a common symptom: all three became exceedingly anxious walking through an empty street or crossing an open space. He called the condition *agoraphobia, agora* being the Greek root for marketplace or place of assembly.

The men also had another symptom in common—a dread of crowded places—but Westphal made less of this. They all had found ways of relieving their anxiety, such as carrying a cane or umbrella, having a few drinks, being with a trusted companion.

The men had the usual physical symptoms of anxiety: palpitations, trembling, feeling of warmth, dry mouth, sweating, and breathlessness. They particularly complained of dizziness. And, in fact, some clinicians felt the dizziness was more important than the anxiety and called the condition "dizziness in public places," perhaps caused by a disorder of the eye muscles.

The term agoraphobia has now survived for a century and at least is easier to pronounce than *Platzschwindel,* the German word for dizziness in public places. Moreover, it is now recognized that dizziness is a minor symptom. Sometimes even a fear of open spaces is not present. Fear of crowded public places is perhaps the single most common complaint by sufferers of agoraphobia.

Diagnostic Criteria

A. The individual has marked fear of, and thus avoids being alone or in public places from which escape might be difficult or help not available in case of sudden incapacitation, e.g., crowds, tunnels, bridges, public transportation.
B. There is increasing constriction of normal activities until the fears or avoidance behavior dominate the individual's life.
C. Not due to a major depressive episode, obsessive compulsive disorder, paranoid personality disorder, or schizophrenia.

Elaboration

Agoraphobia never refers to a single complaint. It refers to a cluster of complaints. The fear of crowded places and, somewhat less often, fear of open spaces are the main complaints but the patient with agoraphobia also dreads some or all of the following:

1. *Public transportation:* trains, buses, subway trains, planes. When crowded, these vehicles become intolerable. Waiting in a line is almost as bad, whether for a bus or a movie.
2. *Other confined places:* tunnels, bridges, elevators, the hairdresser's chair, the dentist's chair, and the barber's chair. (Agoraphobia has been called the "barber's chair syndrome.") These fears belong to the category of "claustrophobia," but most people with claustrophobia have only a single phobia and are not agoraphobics.
3. *Being home alone.* Some agoraphobics require constant companionship, to the despair of friends, neighbors, and family.
4. *Being far away from home* or in places where help cannot be readily obtained if needed. The agoraphobic is sometimes comforted just knowing there is a police officer or a doctor somewhere nearby.

Agoraphobics usually have difficulty explaining why they are afraid. Many say they are afraid of fainting, having a heart attack, dying among strangers. They fear becoming insane. They fear losing control in some manner: screaming or attacking someone (perhaps sexually) or otherwise attracting unwanted attention. The fears are groundless. People with illnesses that may cause fainting, heart attacks, or death almost never have agoraphobia. Agoraphobics do not become insane and do not lose control of themselves in public places.

Agoraphobics have multiple phobias, which is usually not true of individuals with simple or social phobias. In addition, agoraphobia involves more features than phobias. Even when not in clearly defined situations that cause intense fear, agoraphobics still tend to be anxious a good deal of the time. They are subject to depression, especially when thinking about their fears and how their life has been affected by them. In fact, depression is sufficiently common and severe that some authorities believe agoraphobia is a form of depression.

Agoraphobia is the most disabling of the phobic disorders and the hardest to treat. When severe, it can be as disabling as the most crippling forms of schizophrenia. Although agoraphobics almost never require chronic hospitalization, they are sometimes unable to leave home for months or years at a time. The more extreme cases may confine themselves to a single room or spend most of their time in bed. The term "housebound housewife" usually refers to agoraphobics.

Even milder cases involve restrictions in social functioning. Victims are unable to visit friends and neighbors, or go on family outings. They recruit other people to do their shopping and take their children to school. They postpone seeing the dentist and may cut their hair themselves.

Surprisingly, family life is not as damaged by this as one might expect. Spouses will complain about their agoraphobic mate but their marriages apparently survive as well as most. The children tend to be sympathetic and do everything they can to help. The mental health of the children does not seem to suffer.

Agoraphobics are amazingly inventive in discovering ways to mitigate their anxiety so they can maintain at least some social functioning. As Westphal reported, sometimes just carrying an umbrella helps. Other inanimate objects reported to provide relief include canes, shopping baskets on wheels, a bicycle pushed down the street, a folded newspaper carried under the arm. One agoraphobic found relief by loosening his belt. Another said that sucking a piece of candy gave comfort. Agoraphobics almost always are more comfortable in dark places than in sunlight. Some wear dark glasses. They feel better when it is raining. Others can confront crowds if they have a bottle of ammonia or tranquilizers in their pockets (rarely used). If they go to theatres at all, they find an aisle seat near the back to make a fast getaway if necessary.

Studies by Isaac Marks and his colleagues at The Maudsley Hospital in London have uncovered these and many other strategies agoraphobics have discovered to relieve their fears. To the above list Marks adds the following:

> Deserted streets and vehicles are much preferred. Trains are easier to go on if they stop frequently at stations, and if they have a corridor and a toilet. Some journeys are easier if they pass the home of a friend, or a doctor, or a police station, when the patient feels that help is at hand if they get panic-stricken. In such instances if patients know the friend or doctor is not at home their journey becomes more difficult. It is the *possibility* of aid which helps them in their acute anticipatory anxiety before the journey. One patient was able to go on a particular bus route because it passed a police station outside which she would sit if the tension got too much for her. Agoraphobics usually find it easier to travel by car than any other means and may comfortably drive themselves many miles even though they cannot stay on a bus for one stop. Patients with cars may be able to hide their problem for many years.

Everyone agrees, however, that the most reliable fear-reducer is a trusted companion. Many agoraphobics only venture out of the house when accompanied by someone they know and trust: a husband, friend, child, or even a dog.

A newspaper writer in England compiled many of these features into a composite figure called Aggie Phobie: a woman walking at night up a dark alley in the rain while wearing dark glasses, sucking sweets vigorously in her mouth, with one hand holding a dog on a leash, the other trundling a shopping basket on wheels.

The symptoms of agoraphobia tend to wax and wane over the years, waning in part because of the victims' heroic struggle against their fears. Nowhere is this better described than in the following account by a woman writing in *The Lancet*:

> For three years I had been unable to make a train journey alone. I now felt it was essential to my self-esteem to do so successfully. I arranged the journey carefully from one place of safety to another, had all my terrors beforehand, and travelled as if under light anesthesia. I cannot say I lost my fears as a result, but I realized I could do what I had been unable to do.

Now, like others who are disabled, I have my methods. The essentials are my few safety depots—people or places. The safety radius from them grows longer and longer. I am still claustrophobic; that rules out underground trains for me, and I use the district railway. I find it difficult to meet relations and childhood friends, and to visit places where I lived or worked when I was very ill. But I have learned to make short visits to give me a sense of achievement and to follow them when I am ready for it by a longer visit. Both people and places are shrinking to their normal size. Depression usually returns about a week before menstruation, and I have learned to remind myself that life will look different when my period begins. . . . I am also learning that it is permissible to admit to anxiety about things I have always sternly told myself are trifles to be ignored. Many of them, I find, are common fears.

If I am fearful of going anywhere strange to meet my friends I invite them home instead, or meet them at a familiar restaurant. . . . Strangers, too, can be more helpful than they know, and I have used them deliberately; a cheerful bus conductor, a kindly shop assistant, can help me to calm a mounting panic and bring the world into focus again. If I have something difficult to do—to make a journey alone, to sit trapped under the dryer in a hairdresser's or to make a public speech—I know I shall be depressed and acutely afraid beforehand. When the time comes I fortify myself by recalling my past victories, remind myself that I can only die once and that it probably won't be so bad as this. The actual experience now is not much worse than severe stage fright and if someone sees me to the wings I totter on. Surprisingly, no one seems to notice. . . .

I dare not accept my sickness—fear—because it never stays arrested. My very safety devices become distorted and grow into symptoms themselves. I must therefore, as I go along, break down the aids I built up; otherwise the habits of response to fear, or avoidance of occasions of fear, can be as inhibiting as the fear itself.

"It is hard for a free fish to understand what is happening to a hooked one," writes Karl Menninger. Clearly, we may not even know when someone is hooked. How many umbrellas are safety devices—and not against the weather? What are sunglasses shading? Sunlight or fear? How often is the drunk on the street there only because, sober, he would never venture from the house? Nobody knows how many agoraphobics are among us in various disguises, but the number may be larger than anyone suspects.

Westphal in his 1871 paper mentioned that alcohol emboldened his three male agoraphobics to venture out into open spaces. Other agoraphobics have discovered the same thing, and that tranquilizers also help. Are agoraphobics susceptible to alcoholism and drug abuse? The evidence is conflicting. In two studies, nearly one-third of a group of alcoholics gave a history of agoraphobia or social phobia. (There is no comparable study of agoraphobia among drug addicts.) A study of agoraphobics, on the other hand, showed a relatively low rate of alcoholism and drug addiction (5–10%). Since alcohol unquestionably reduces anxiety, it certainly would make sense if anxious people, including agoraphobics (probably the most anxious of all), resorted to alcohol for relief. It would make further sense that, having resorted to alcohol on repeated occasions, dependence on alcohol would occur.

In fact, most long-term studies do not indicate that alcoholics were unusually anxious as children or adolescents. However, phobia, as a distinct clinical disorder, must be distinguished from anxiety, and has been little studied in connection with alcoholism. Other studies are needed.

More consideration has been given to the possibility that agoraphobia is related to clinical depressions. In common with clinical depressions, agoraphobia often begins "out of the blue": the first symptoms occur for no apparent reason. Also, agoraphobics often are depressed and their depressions may resemble the depression seen, for example, in manic-depressive disease. Conversely, symptoms associated with agoraphobia are not uncommon in individuals with depression. Therefore, the issue of whether agoraphobia is a form of depression has not been resolved. A recent study suggested they are not the same illness. A laboratory test called the dexamethasone suppression test, which is sometimes useful in diagnosing depression, has been reported negative in agoraphobia.

Agoraphobia does not always begin out of the blue. Sometimes it is preceded by a physical illness or stress such as an important examination or marital conflict. The symptoms may come on all at once or gradually. Some agoraphobics have no symptoms at all until one day they are standing at a bus stop or shopping in a large store when they suddenly panic, rush home, stay indoors, and for years later avoid a variety of public places, including bus stops and large stores. Other agoraphobics are not able to say exactly when their illness began.

Once the illness has developed, victims often have trouble deciding which is worse: the anxiety that occurs in the situation itself (a store, a bus) or the anticipation of becoming anxious. The agoraphobic may start feeling anxious the moment he or she awakens, thinking about the day ahead. Some find the anticipatory anxiety is greater than the anxiety in the feared situation—an observation with treatment implications (Chapter 18).

How common is agoraphobia? With recent publicity about agoraphobia, more and more patients are becoming visible in one way or another. Little attention was paid to the disorder until recent years—one reason perhaps being that phobics, and especially agoraphobics, tend to be secretive about their symptoms. As for physicians, many have been unaware that the disorder existed. This has now changed. A recent study by the National Institute of Mental Health found that one adult in 20 suffers from agoraphobia.

Agoraphobia usually begins in the mid or late twenties, almost never occurring before 18 or after 35. At least two-thirds of agoraphobics are women. Of the three types of phobia, agoraphobia has the latest age of onset, with simple phobia usually occurring in childhood and social phobia in adolescence.

What happens to agoraphobics? If they do not receive treatment, about one-half of the mild or moderately disabled agoraphobics will recover or substantially improve five to ten years after the illness begins. The severely disabled housebound patient may remain that way indefinitely. Some cases are short-lived but nobody knows how many. Most clinicians would agree with Isaac Marks that if the symptoms of agoraphobia persist for a year or longer they are almost certain to last much longer.

17

Separation Anxiety Disorder (School Phobia)

> A diller, a dollar
> A ten o'clock scholar;
> What makes you come so soon?
> You used to come at ten o'clock,
> And now you come at noon.
>
> Mother Goose

Separation anxiety disorder—better known as school phobia—is an exaggerated fear of attending school, usually related to anxiety about leaving home. It occurs in children of all ages, peaking around 11 or 12. Both sexes are affected, girls perhaps somewhat more than boys. It is encountered in children of every social class and with the full range of academic abilities.

Diagnostic Criteria

A. Excessive anxiety concerning separation from those to whom the child is attached, as manifested by at least three of the following:

1. Unrealistic worry about possible harm befalling major attachment figures or fear that they will leave and will not return
2. Unrealistic worry that an untoward calamitous event will separate the child from a major attachment figure, e.g., the child will be lost, kidnapped, killed, or be the victim of an accident
3. Persistent reluctance or refusal to go to school in order to stay with major attachment figures or at home
4. Persistent reluctance or refusal to go to sleep without being next to a major attachment figure or to go to sleep away from home

5. Persistent avoidance of being alone in the home and emotional upset if unable to follow the major attachment figure around the home
6. Repeated nightmares involving theme of separation
7. Complaints of physical symptoms on school days, e.g., stomachaches, headaches, nausea, vomiting
8. Signs of excessive distress upon separation, or when anticipating separation, from major attachment figures, e.g., temper tantrums or crying, pleading with parents not to leave (for children below the age of six, the distress must be of panic proportions)
9. Social withdrawal, apathy, sadness or difficulty concentrating on work or play when not with a major attachment figure

B. Duration of disturbance of at least two weeks.
C. Not due to a pervasive developmental disorder, schizophrenia, or any other psychotic disorder.

Elaboration

The first sign of a school phobia is often a physical complaint: stomachache, sore throat, headache. The child complains of being too ill to attend school and when the mother finally gives in and says he or she can stay home, the complaint almost magically disappears. If the mother does not give in, the illness may persist until the school nurse sends the child home. The symptoms go away only to return on the next morning of a school day.

After this continues for a time and the pediatrician finds nothing wrong, the diagnosis becomes "school phobia" and the school counselor takes over. What might the counselor learn?

School phobia can almost always be traced to one of two causes: a fear connected with school or, more frequently, a fear of leaving home. School can be frightening in many ways. The teacher may tend to be frightening. Gym may be frightening, particularly undressing in front of others. Reciting in class may be frightening, often associated with a fear of fainting. Children may be teased about their looks or dress and fear going to school because of this (mothers who dress their children oddly increase this risk). Children may develop toilet phobias and want to stay home where there is privacy. Finally, any change from the status quo, particularly for the timid child, can lead to phobia, the most common examples being changes to new schools with new faces and challenges.

Much more frequently, however, what seems to be school phobia turns out to be a fear of leaving home. (This happens so often that DSM-III has renamed school phobia "Separation Anxiety Disorder.") There has been a death in the family, or separation or divorce, or sometimes a new baby. Or maybe the parents argue a lot and there is financial trouble. Maybe none of these things has happened but the child is overdependent on the mother. This sometimes shows up when the child, forced to go to school, insists on repeatedly calling the mother on the telephone.

Sometimes it is hard to distinguish school fear from separation fear, and here is how one counselor handled the problem:

> Billy, 11 years old, had been absent from school for seven weeks. After being absent for three weeks because of illness, he had developed a phobia towards school. He became anxious, unable to eat, and complained of chest pains. Neither punishment nor bribes led to a return to school. He said he was unaware of reasons for his fear of school.
>
> Billy was the youngest of two children, and was described as submissive, tense, perfectionistic, and very attached to his family. He had many physical illnesses and his mother was overprotective. Increased school competition and many absences led to poor grades. He was teased by peers because of academic difficulties and his weak and sickly appearance.

Over many hours, the counselor conducted what is described as a "behavioral assessment procedure." This meant asking Billy to visualize and describe a typical school day, noting indications of anxiety. These included flushing of the skin, body movements, muscular tension, vocal tremors, and tears. Mathematics and literature both were sources of anxiety. Being called on, unable to answer questions, and being teased evoked intense anxiety. Billy showed no signs of anxiety when asked to visualize getting up in the morning, having breakfast, and preparing to leave home. The counselor concluded that the school phobia was indeed a phobia of schools.

Of the next 100 children who move up to a new school year, two or three will develop phobias related to school sufficient to concern the parents and teachers and, ultimately, interfere with the child's school performance. What is there about these two or three children that distinguishes them from the unaffected majority?

In studies, one encounters numerous examples of children, mothers, and families "typically" associated with school phobia. It is said, for example, that children with school phobia commonly overvalue themselves and their achievements. When their own estimate of themselves is threatened, they become anxious and withdraw from competition, often seeking closer contact with the mother. Their estimate of themselves is not securely held, and consequently they are sensitive to threat, such as change to another class or new school, return to school after an illness, a minor episode that leads to embarrassment, and actual or fantasied academic or social failure.

The "typical" mother develops unusually close dependence on and from her children as a compensation for her unsatisfactory marital or other relationships. The mothers themselves often have a history of an unhappy relationship with their own parents. Such mothers are reluctant to leave their children in school and will tell the teacher, in the child's presence, "You won't be able to get him to leave me."

In the "typical" family one sees the overdependent child and overprotective mother who is overdependent herself on a husband who is impatient with both of them. Usually, school phobia is reported in small families, but in one study the opposite was true: families with the phobic child lived with the grandparents or telephoned them daily.

Because one version of "typical" varies so often from others, I have put the word in skeptical quotes. School phobia develops for reasons that nobody understands and words like "over-dependency" help little in explaining why a small number of children develop the problem and most do not.

School phobia must be distinguished from truancy. Many children do not go to school simply because they prefer not to. They prefer doing something else, video games today being high on the list. Truant children are often delinquent in other ways, getting into trouble with authorities, running away from home, and generally being hard to manage. Most inmates in penitentiaries were truant as children. Very few children phobics end up in penitentiaries.

On extremely rare occasions, the child who avoids school (and usually most other social activities) becomes schizophrenic. They are loners first and later frankly psychotic. Children with school phobia usually have friends, although sometimes their school phobia will expand to include a phobia toward parties and visits to relatives or other children's homes.

Children can become depressed like adults. When school refusal is accompanied by irritability, crying, insomnia, and concern about death, depression is probably more likely than simple school phobia.

How do children with school phobia get along later in life? Most studies show they do about as well as other children, perhaps being somewhat on the sensitive, cautious side, but suffering from no diagnosable mental illness. This benign prospect applies more to cases where the phobia developed early in life rather than in adolescence. When school phobia first develops in adolescence, the likelihood is increased that school phobia is not the only explanation, but that schizophrenia, depression, or criminality will ultimately be the outcome.

Most experts agree that the best treatment for school phobia consists of getting the child back in school as early as possible, by whatever means. Determining the source of the phobia (e.g., teasing) will of course be useful if it can be removed. Even when the source is not clear, however, a firm mother and equally firm and understanding teacher can gradually cajole a child into attending class, sometimes by having him sit for a few days in the library or Principal's office, attending class briefly and then for increasingly longer periods, and sometimes with the mother attending class with him for a few days. Everyone should evince confidence that the child will naturally return to school and the only question is when. Failure to return to school in a reasonable time leads to a worsening of the phobia, as the child falls steadily behind in his school work, builds up resentment, and feels even more different from his classmates than in the beginning.

There is evidence that antidepressant drugs are helpful for school phobia, as will be discussed in the next chapter.

18

Treatment for Phobias

> One can hardly ever master a phobia if one waits til the patient lets the analysis influence him to give it up. . . . One succeeds only when one can induce them . . . to go about it alone and to struggle with their anxiety while they make the attempt.
>
> Sigmund Freud

Psychiatrists and psychologists have been treating phobias for the past hundred years but until recently there was little evidence that the treatments were effective. Traditional psychotherapy—ranging from "supportive" therapy to psychoanalysis—did not seem to help much, whether the patients were seen individually or in groups. One study found that patients psychoanalyzed for 10 or 15 years showed slight or no improvement. Another found that only 13% of a mixed group of phobics were free of symptoms after intensive psychotherapy. The recovery rate for untreated patients is usually higher than this.

About 25 years ago, this all changed. The watershed year was 1958, when Joseph Wolpe proposed a new treatment for phobia in a book called *Psychotherapy by Reciprocal Inhibition*. It was not entirely new. Behind the treatment was an old theory—learning theory—dating back to Pavlov and the behaviorists of the 1920s. Its message was not new either. Years before, Freud had said that psychoanalysis was futile unless phobics "struggle with their anxiety" and confront what they fear. It *was* new in that Wolpe said it did not matter how the phobia came about, or whether the patient understood the origins of the phobia. The phobia could be attacked head-on as a learned condition that could be unlearned. He gave specific instructions for the unlearning part. He also claimed great success for his treatment: a 95% recovery rate!

Later there was some disillusionment about the Wolpean "behavior therapy"; nobody, including Wolpe, ever got so high a recovery rate again.

Enthusiasm, however, for other kinds of behavior therapy has grown steadily. Imaginative, optimistic, always trying something different, a generation of behavior therapists has become the dominant force in the treatment of phobias. Today, the person who seeks help for a phobia will probably receive some type of behavior therapy.

It may be with or without drugs. There is rather convincing evidence that two types of drugs usually prescribed for depression relieve panic attacks. Many therapists treat phobias with behavior therapy *and* drugs.

For these reasons, most of what follows concerns behavior therapy and drugs.

Behavior Therapy

There are many kinds of behavior therapy, but they all have a common goal: reduction of anxiety by exposure to the phobic situation. They differ only in how this is accomplished.

The exposure can be "graded" or "ungraded." It can occur in real life or in the person's imagination.

Graded exposure is based on the artichoke principle: you take it leaf by leaf. It is also the behaviorist equivalent of painless dentistry. The only distress the patient may feel is boredom; graded-exposure behavior therapy á la Wolpe can be excruciatingly tedious, for the therapist if not the patient (probably explaining why so many therapists favor other procedures).

Here is Wolpe's approach:

1. Identify the phobia.
2. Grade the phobia in its various aspects from least fearful to most fearful (called "constructing a hierarchy").
3. Instruct the patient to imagine, or visualize, a situation involving the least fearful aspect of the phobia.
4. Combine the visualization with a pleasant experience—usually muscle relaxation—to counteract the mild fear created.
5. When the patient can visualize a mildly fearful situation without feeling anxious, "move up" the hierarchy to a slightly more anxiety-provoking imaginary situation. Continue this, step by step, until the patient can visualize the most fear-arousing situation in the hierarchy and not experience fear.

TREATMENT FOR PHOBIAS

At this point is the patient cured? Perhaps not altogether. Neo-Wolpeans insist he must now go out into the world and practice confronting real-life phobic situations for the treatment to be fully effective. Here is a typical hierarchy for a snake phobia, rating anxiety on a 6-point scale (6 being most anxious).

1. Seeing a picture of a snake: 1-plus.
2. Twenty feet from a small garden snake: 2-plus.
3. Twenty feet from a boa constrictor: 3-plus.
4. Five feet from any snake: 4-plus.
5. Touching a snake with a gloved hand: 5-plus.
6. Holding snake in lap: 6-plus.

The patient is first taught a method of "deep relaxation"—a pleasurable state. He then relaxes while visualizing a picture of a snake. If no anxiety occurs, he holds up a finger, and the therapist tells him to visualize step 2. No anxiety? Step 3. No anxiety? Step 4. Finally the patient visualizes a snake in his lap with no anxiety—and no phobia.

Whenever he moves up a step, relaxed but feeling anxious, he is told to go back down a step and try again. Many exhausting sessions may be required to reach step 6, but infinitely fewer than in psychoanalysis—at a fraction of the cost.

The process of moving up the hierarchy is called "systematic desensitization." The method works, Wolpe believed, because of "reciprocal inhibition," referring to the fact that two incompatible emotions cannot be experienced simultaneously.

Two such emotions are fear and pleasure. Relaxation is pleasurable. There are various methods for becoming relaxed. Wolpe favored the method of Edmund Jacobsen, consisting of first tensing a muscle group and then releasing the tension beyond the normal resting tension. When all muscle groups have been tensed and relaxed, a pleasurable state of deep relaxation results.

Once a person learns to relax deeply, he can then expose himself to mildly fearful stimuli—in his imagination or real life—and not experience the incompatible emotion of fear. Having conquered mild fear, he can practice relaxing in situations involving increasingly greater fear. The sweetly-smiling, drowsy-looking man you saw on the park bench holding the boa constrictor in his lap is likely the satisfied customer of a systematic desensitizer.

There are other pleasures than relaxation. Sometimes a tranquilizer or sedative like sodium amytal is given to phobics moving up a hierarchy. Sometimes relaxation is helped along by hypnosis. Success with the Wolpe method obviously requires a patient with a vivid imagination. Not everybody has one. Some patients are too anxious to visualize a scary scene. Some find the whole business a bit silly and burst out laughing when told to visualize a boa constrictor. This annoys the therapist and the treatment comes to a halt, phobia intact.

This—plus some doubts about the effectiveness of the method—led behavior therapists to try something more daring: expose the patient to *real* snakes (or whatever is feared). Again, a hierarchy is constructed. The patient moves up the hierarchy at his own pace, keeping anxiety at a comfortable level while gradually increasing exposure to the feared object or situation. The difference is that the snakes, the crowds, and the elevators are all real.

Real-life exposure has two drawbacks: (1) some patients refuse to cooperate because of anticipatory anxiety, and (2) real-life hierarchies are hard to construct in real life. Snakes are fairly easily obtained, although perhaps not boa constrictors. Fire escapes are handy for height phobias—steps 1, 2, and 3 become literally steps 1, 2, and 3. But what about thunderstorm phobias? Any loud, booming noise might serve for steps 1, or 2, but a thunderstorm is needed for step 3. Scheduling a therapy session to coincide with a thunderstorm calls for skills beyond those usually possessed by behavior therapists.

Which raises the question: does the therapist have to be present while the patient moves up a hierarchy? The answer is: it helps. As you may recall from earlier chapters, no *soteria*—fear dispeller—is more effective than a trusted companion. A *paid* trusted companion is perhaps even better, and so therapists take full advantage of their *soteria* role by actively engaging the patient in something called "participant modeling."

This wrinkle on Wolpean therapy is based on the principle that nothing succeeds like success—even someone else's success. Before the patient moves up a step in the hierarchy, he watches the therapist demonstrate that it can be done. The therapist moves closer to the snake; the patient then moves closer. The therapist touches the snake; the patient touches it. Courage is contagious and seeing others do something we fear makes it easier for us to do it ourselves.

The therapist helps the phobic patient climb the hierarchy in various other ways. The patient may be instructed to place his hand on the therapist's hand while the latter touches the snake. The therapist may lavishly compliment the patient for every small success while withholding criticism. He may keep a log of the patient's progress: the length of exposure, the closeness of contact, the number of hierarchy items completed. Whenever the patient exceeds the day's goal for items, he is told, "That's great!" (Therapists call this "reinforced practice.") He may improvise for each patient a number of "performance aids"—for example, in animal phobias, protective clothing for the patient and a leash for the animal.

Overinvolvement by the therapist poses risks. The patient may do well as long as the therapist is there but be as phobic as ever when the therapist is absent. The goal of behavior therapy, like all therapy, is to eliminate the therapist. All therapies must be tested by the standard: does the new learning transfer to real life? This is why therapists emphasize "mastery practice" between sessions.

So much for the graded approach to phobias. It works sometimes and sometimes does not. It works best with simple limited phobias. It is usually found somewhat superior to other psychotherapies, but not always. (Most people with snake phobias do not seek psychotherapy.) At least the pioneers of behavior therapy have been scientifically minded—they have sought proof that their treatments work. Single case histories have been played down. As a result, there exists a body of evidence bearing on the effectiveness of the various behavior treatments.

One drawback in these studies is that most have been conducted with nonpatients—typically college girls with snake phobias or fear of heights. These are problems that in themselves rarely lead to psychiatric consultation.

In any case, it is clear that systematic desensitization, whether in real life or in the person's imagination, is not the panacea its originators had hoped it would be. Behaviorists looked for other approaches and the one that has caught on most in recent years was something called "flooding." This involves *ungraded exposure* to the phobic stimulus.

The person is afraid of heights? Take him to the roof of the Empire State building and have him lean over the guard rail. He will be terrified of course. Terror is not an emotion that can be sustained indefinitely and, little by little, he will find his anxiety diminishing. In other words, move directly to step 6 and be done with it. "People don't die of anxiety," flooders say, then warn against flooding anyone who has heart disease or other illnesses where anxiety possibly can kill.

The treatment is beautifully simple. A phobia can be cured in a single session, although usually it takes several. The trick is to keep the phobic in a state of maximal panic for as long as necessary for the panic level to drop. If the person is removed from the phobic situation before the panic drops, he will just be worse for the experience. How do you get people to walk into burning houses when they are afraid of fire? Jump off bridges when they are afraid of both heights *and* water? Flooders answer that you just do it. You stay with the patient. You encourage him, you praise him, remind him that he promised he would go through it. You do not resort to physical means but to almost everything else.

Repeated trials of prolonged exposure to real-life stimuli of maximal phobic intensity do indeed produce results, as would have been predicted by learning theory which holds that phobic anxiety is sustained by repeated avoidance of anxiety-eliciting stimuli. The trick is to find people who are willing to go through it. Flooding works with animals: a strongly conditioned fear response can be extinguished by physically restraining the animal in a fear-eliciting situation. People are not so restrainable.

Then, too, there is the problem of eliminating the therapist. Everyone agrees that phobics cannot go through life with a therapist at their side, or even a friend or relative. The cure of phobia comes when the phobic can confront the feared situation and "go about it alone," as Freud said.

The good therapist actively works at self-elimination.

> A 22-year-old man refused to ride a bus by himself. The therapist agreed to accompany him during one session but suggested they sit in different sections of the bus rather than next to each other. The patient agreed to the compromise and was able to tolerate the anxiety elicited during the one-hour bus trip. In the following session, the patient successfully traveled by bus alone.

Real-life flooding is often hard to arrange and therapists fall back on imaginary flooding. After learning what frightens the patient the most, the therapist asks the patient to visualize his most frightening scene and helps by describing the scene. The therapist watches for signs of anxiety in the patient and repeats the scenes that produce the most anxiety. As with real-life flooding, he keeps it up until the anxiety level falls.

Some therapists embellish flooding by describing exaggerated catastrophic scenes to the patient. The patient is afraid to drive and is asked to imagine that he is driving a school bus that plunges over a cliff with everyone killed. Frightened of speaking out at a small meeting, he imagines that he is the keynote speaker at a convention in Madison Square Garden. This is called *implosion therapy*. It is not widely used, being as disliked by most therapists as by patients.

What works best? There are studies showing that flooding is better than systematic desensitization; studies showing that systematic desensitization is better than flooding; and studies showing both with the same effect or no effect. What almost all studies do show is that real-life exposure is superior to imaginary exposure, whether graded or ungraded. The combination of real-life exposure with "participant modeling" and "reinforced practice" seems particularly effective.

Drugs for Phobias

In the 1960s, as behavior therapy began to flourish, a group of psychiatrists in England and another in New York began studying the effectiveness of drugs for phobia. Phobics from time immemorial had taken drugs: bromides, chloral hydrate, barbiturates, derivatives of the poppy plant and hemp, and, of course, alcohol. All relieve anxiety, but briefly; all produce hangover or withdrawal, and all are potentially addictive. Responsible doctors in the twentieth century prescribe them as little as possible. Then, in the 1950s, three new classes of drugs—synthetic and new under the sun—came on the market.

One group was for psychoses. Another was for anxiety. The third was for depression.

The new behavior therapies helped some phobics, but not all; the sickest phobic, the agoraphobic, was particularly unresponsive to behavior therapy. So the new drugs were tried, and here are the results.

Tranquilizers

Drugs for psychoses are called "major tranquilizers" or neuroleptics; Thorazine was the first one introduced. There is no evidence these drugs help phobias.

Drugs for anxiety are called "minor tranquilizers" or anxiolytics, usually referring to the benzodiazepine class of drugs; Librium and Valium are examples. Since phobias are anxiety states *par excellence*, one would expect that if any drug helped phobias, it would be a drug that relieves anxiety, as Librium and Valium clearly do.

These drugs are indeed widely prescribed for phobias, but, surprisingly, their effectiveness has not been evaluated. They are widely believed ineffective for agoraphobia, but may help simple and social phobias by lowering anticipatory anxiety and giving the "chemical courage" required to confront a feared situation.

In the case of flying phobias, the airport bar has performed this service through the years. It is not known how many people today board planes fortified with a minor tranquilizer (sometimes combined with alcohol), but one suspects the number is sizable. One would hope that if a Valiumized phobic boarded enough airplanes, eventually he could board planes anxiety-free *without* Valium. It is not clear that this happens. In animal studies, sedatives and alcohol suppress phobia-like avoidance behavior, but only as long as the animal remains drugged or intoxicated; when the drug is withdrawn, the avoidance behavior returns. The situation for humans may not be quite so pessimistic. Many people give tranquilizers credit for overcoming stage fright and eventually perform without them (often, of course, keeping a capsule or two in a jacket pocket just in case).

The minor tranquilizers relieve moderate degrees of anxiety but apparently do not prevent panic attacks. Xanax (alprazolam), a "second-generation" minor tranquilizer, may be an exception. High doses apparently prevent panic attacks as well as relieve anticipatory anxiety.

Patients with simple and social phobias usually do not need to take minor tranquilizers on a regular daily basis. They are not generally anxious people, and there is evidence that people who are not generally anxious *become* anxious if they take tranquilizers. The tranquilizer should be taken, if at all, shortly before the anxiety-producing situation. Sometimes, as noted, just carrying them in one's pocket relieves anxiety sufficiently for the person to confront the phobic situation, which is what really counts in achieving a lasting recovery from a phobia.

Valium and other minor tranquilizers are the most widely prescribed medications in the world, and, compared to barbiturates and other sedatives, are relatively safe. It is virtually impossible to kill oneself by taking an overdose of the drugs and relatively few people become physically dependent on them, meaning they take high doses and have serious withdrawal symptoms, like seizures, when they stop taking the drug. There is some evidence that *long-term* use of minor tranquilizers, even in the doses normally prescribed, results in some withdrawal effects if the drug is suddenly discontinued, but this is not definitely established. The drugs tend to lose their effect after several weeks of taking them daily, and have another disadvantage: some are slowly eliminated from the body. If taken regularly, they accumulate in the body to rather high levels. This may result in the person being constantly drowsy in the daytime, with the implications this has for driving and handling machinery. However, short-acting tranquilizers are now available and are probably preferable for phobias.

For these reasons, the drugs should only be prescribed by physicians who are thoroughly familiar with them, and today this usually means a psychiatrist.

Drugs for Depression

Two new groups of drugs for depression were introduced in the 1950s and 1960s. One group consists of something called MAO inhibitors and the others are called tricyclics. Phenelzine is a member of the first group and imipramine belongs to the second. By constantly tinkering with the molecules, drug manufacturers have brought on the market more than 20 drugs in the two groups. There is little evidence that one drug is superior to another in each group, although the side effects differ.

In 1962, William Sergeant in England reported that 60 patients with anxiety states, including phobias, responded well to MAO inhibitors (sometimes given with a minor tranquilizer). Nearly a dozen studies over the past 20 years have confirmed the finding. They all show that MAO inhibitors—particularly phenelzine—provide some relief for phobias, reducing the anxiety if not the avoidance behavior. Although the drugs are usually prescribed for depression, depressive symptoms may be absent and the drugs will still alleviate phobic anxiety.

The drugs take about a month to work and then usually continue to be effective until stopped. When stopped, the phobias almost always return. In other words, in learning theory terms, the "learned" fear does not become extinguished or "unlearned" through repeated exposure to the phobic stimulus when the exposure is facilitated by an MAO inhibitor. The almost inevitable relapse when the drug is discontinued provides strong evidence that the drug actually works. On the other hand,

MAO inhibitors are potentially dangerous and physicians balk at prescribing them indefinitely.

The patients similarly balk at taking them indefinitely. When given the drug they are also given a long list of foods, drinks, and other medications to be avoided, including aged cheeses, wines, beer, pickled herring, broad bean pods, yogurt, yeast, hayfever pills, and other antidepressant medications. Most patients can do without broad bean pods, but they face life denied cheese and beer with little enthusiasm. Consequently, most doctors do not prescribe an MAO inhibitor until they have tried the tricyclic class of antidepressants first.

It was also in 1962 when Donald Klein and his group in New York reported that a tricyclic antidepressant, imipramine, was an effective treatment for agoraphobia. In studies over the next two decades, they confirmed their original finding and became increasingly specific about the drug's mode of action.

Klein believes that agoraphobia has three parts: (1) spontaneous panic attacks, (2) anticipatory anxiety, and (3) avoidance behavior. Imipramine abolishes the panic attack but has little effect on the anticipatory anxiety or avoidance behavior. He recommends combining imipramine with psychotherapy. In Klein's studies, supportive psychotherapy does as well or better than behavior therapy in preventing anticipatory anxiety and avoidances. ("Supportive" means giving encouragement, reassurance, and sometimes advice, without delving deeper.) Most of Klein's patients who responded to imipramine had a history of receiving psychoanalysis or other intensive psychotherapy, and neither seemed to help. A combination of imipramine and supportive psychotherapy not only worked but was economical.

Imipramine did not relieve simple phobias. Klein's explanation was that simple phobias are usually not accompanied by a panic attack. Many students of agoraphobia, starting with Freud, believe that a spontaneous panic attack is the initial manifestation of agoraphobia, and what agoraphobics fear most, when they stay at home, avoiding travel and crowds, is repetition of the initial panic. The panic is "spontaneous" in that there is usually no apparent connection with a traumatic event. People with simple phobias more often associate the beginning of their phobia with a trauma. Agoraphobics often have a history of school phobia and separation anxiety associated with real or threatened loss of a parent, but their first panic attack typically comes "out of the blue."

Since imipramine relieves the panic attacks of agoraphobia, and since agoraphobics have a history of school phobia, Klein reasoned that imipramine might also help in school phobia. His group did a study and, sure enough, imipramine had a potent effect on school phobia. It is now probably the best treatment for this problem, when persuasion doesn't work.

A high relapse rate follows the discontinuation of MAO inhibitors, and the same is true of imipramine. Even after patients have taken imipramine for six months or longer, one-third or more resume having panic attacks immediately upon cessation of the drug. In short, there is little or no transfer from drug to nondrug state.

This might have been predicted because of a phenomenon called "state-dependent learning." People and animals trained to perform tasks under the influence of mind-altering drugs perform well as long as they are under their influence and do poorly when the drug is withdrawn. Their performance is "state-dependent," the state being one of intoxication on a particular drug. New behavior learned in a drug state may not transfer to a nondrug state, and this is one of the disadvantages of giving mood-altering drugs, particularly drugs with side effects or long-term hazards.

Imipramine and other tricyclic drugs have many side effects. They range from dry mouth and blurred vision to increased appetite and impotence. Not everyone experiences the side effects and they tend to go away as the drug is taken over a period of time. However, phobic patients seem especially sensitive to the side effects of MAO inhibitors and tricyclic drugs, explaining why there is a high drop-out rate in studies of these drugs. Phobics are particularly bothered by the stimulant effects of imipramine. The drug makes them jittery and causes insomnia, rather like Dexadrine. Other drugs in the tricyclic family have sedative rather than stimulant effects, and there is limited evidence that these drugs may be as effective as imipramine in suppressing panic attacks. Also, adjusting dosage often reduces side effects.

Klein noted that not only did imipramine suppress panic attacks but improved the "quality of life" of his patients. They got along better with friends and relatives, perhaps because some phobias are as troublesome for friends and relatives as for the victim. He noticed something observed by the behavior therapist: patients relieved of phobias do not develop "substitute" symptoms. Psychoanalytic theory holds that phobias are symptoms of unconscious conflicts, that if one symptom goes away another appears in its place unless the conflicts are resolved. Studies of phobias over the past 20 years tend to refute this idea.

Like behavior therapy, drug therapy of phobias was ushered in with much enthusiasm. This has waned somewhat over the past 20 years, without evaporating. Drugs help; hardly anybody questions this any longer. But they do not help everyone and drugs alone rarely cure phobias. Klein recommends a combination of drugs and supportive therapy. In any case, a combination of something is usually required, and no combination represents a truly definitive treatment, in the sense that penicillin is a definitive treatment for syphilis. People tend to get over phobias. The response to placebo in all studies of phobia is remarkable; at least one-third of patients improve on a sugar pill alone. Good responders to treatment are often the same people likely to recover anyway: people with good general mental health and an illness of short duration.

Still, there has been much progress in the treatment of phobia over the past 20 years and it is not overly optimistic to foresee a time when the cause of phobia will be understood and the treatment as spectacular as penicillin. (At the time of writing, more than 70 new drugs for anxiety and depression are in various stages of development.)

Beta Blockers

Any discussion of drugs for phobias should include propranolol, the most widely prescribed so-called beta blocker. Beta blockers are usually prescribed for high blood pressure and irregular heart action. However, they also are a good remedy for trembling hands, racing pulse, and other physical expressions of anxiety.

William James believed that if you could eliminate the physical expressions of anxiety you would also eliminate anxiety (The James-Lange theory; see Chapter 3). Since nobody's pulse races faster than a phobic's in a phobic situation, and no hands tremble more, it is surprising that beta blockers have not been studied for phobias. One reason they have not been is that they are rather tricky drugs with numerous side effects, and cannot be taken by people with various common diseases. Nevertheless, the drugs are probably as safe as the antidepressants which, to everyone's surprise, relieve panic as well as depression.

As a matter of fact, beta blockers have been studied for one type of phobia: stage fright. Some years ago, clinical pharmacologists of the Royal Free Hospital in London hired Wigmore Hall and engaged 24 string players who had histories of stage fright to perform under the influence of a beta blocker. It worked. According to a trade newspaper, the drug dramatically reduced the effects of stage fright without detriment to technical execution. In fact, teachers, performers, and critics involved in the study noted significant improvement in accuracy, rhythmic stability, and memory among the propranolol users. In the same issue, a violinist called attention to an ethical issue: "Might not use of potent prescription drugs by a performer at an audition give him an unfair edge over the competitor just as it might to the athlete or race horse? Must orchestras be prepared to administer blood and urine tests to audition applicants?"

Why would an orchestral musician have stage fright? It seems to violate a basic principle of learning theory: phobias last only as long as the phobic situation is avoided. How can an orchestral musician avoid orchestras? There is even safety in numbers: if a violinist has not quite mastered a difficult passage in "Thus spake Zarathustra," his discreet miming of the fingering and the bowing would not be noticed, except perhaps by the colleague in the next chair.

This notion overlooks the fact that although the performance life of most orchestral players is corporate and comparatively free of anxiety, many performers regularly take solo roles, without the orchestra, as principals or section leaders. A cellist describes the symptoms he and his colleagues may experience: "That dreaded onset of sweaty palms, racing pulse, trembling hands, dry mouth, labored breathing, nausea and memory loss."

There are more phobias around than most people suspect. There are also many treatments, with more to come.

FOR EXAMPLE, a recent treatment for agoraphobia involved neither behavior therapy nor drugs. Anxiety was reduced simply by improved breathing.

As mentioned earlier, overbreathing or hyperventilation occurs frequently during anxiety states and produces tingling and dizziness. The latter, in turn, makes the person even more anxious, creating a vicious cycle. A group of British researchers decided to train agoraphobics to breathe in a manner to avoid the vicious cycle.

First they found that agoraphobics were indeed subject to hyperventilation. Of 21 people with agoraphobia who were asked to hyperventilate for three minutes, 14 were unable to do so because of the distress produced before the time was up. By comparison, only two of 47 control subjects could not complete the three minutes of hyperventilation.

The agoraphobic patients were then instructed in how to breathe with their diaphragms, keeping their chest walls as immobile as possible—a technique designed to prevent hyperventilation. When reexposed to frightening situations, these individuals experienced far fewer panic attacks than before. Another group of agoraphobic patients who had received behavior therapy, but no retraining in how to breathe, showed a similar degree of improvement for a period of ten weeks. However, at the end of six months, the ones who had learned the new breathing technique were faring better than those who did not.

Reported in *The Lancet* (September 22, 1984), the study illustrates the enormous potential for research in phobias and the other anxiety disorders, and the promise this research holds.

19

Obsessive Compulsive Disorder

> They that live in fear are never free.
> Robert Burton
> *The Anatomy of Melancholy*

Obsessive compulsive disorder is a chronic or recurring illness dominated by obsessions and compulsions. The condition occurs in 1%, or less, of the adult population. Both sexes are affected equally.

Obsessionals are even more secretive about their symptoms than phobics are. If they seek professional help, it is usually because of depression. Sometimes psychiatrists will see them for years without suspecting they have obsessions.

Diagnostic Criteria

A. Either obsessions or compulsions:

- *Obsessions:* recurrent, persistent ideas, thoughts, images, or impulses that are ego-dystonic, i.e., they are not experienced as voluntarily produced, but rather as thoughts that invade consciousness and are experienced as senseless or repugnant. Attempts are made to ignore or suppress them.
- *Compulsions:* repetitive and seemingly purposeful behaviors that are performed according to certain rules or in a stereotyped fashion. The behavior is not an end in itself, but is designed to produce or prevent some future event or situation. However, either the activity is not connected in a realistic way with what it is designed to produce or prevent, or may be clearly excessive. The act is performed with a sense of subjective compulsion coupled with a desire to resist the compulsion (at least initially). The individual generally recognizes the senselessness of the behavior (this

may not be true for young children) and does not derive pleasure from carrying out the activity, although it provides a release of tension.

B. The obsessions or compulsions are a significant source of distress to the individual or interfere with social or role functioning.

C. Not due to another mental disorder, such as Tourette's disorder, schizophrenia, major depression, or organic mental disorder.

Elaboration

Obsessive compulsive disorder almost always begins in adolescence or the early twenties. The symptoms are extremely diverse. A common complaint is the persistent and irrational fear of injuring oneself or another person, often a child or close relative. Fearful of losing control, the patient may develop avoidances or rituals that lead in turn to social incapacity. Perhaps he will refuse to leave the house, or will avoid sharp objects, or wash repeatedly to destroy "germs."

The obsessional may recognize that the thought is illogical, but not always. Sometimes the ideas are not, strictly speaking, illogical (germs do cause disease), and sometimes, even when obviously absurd to others, the ideas are not seen this way by the obsessional. What distinguishes an obsession from a delusion is not so much "insight" (recognizing the idea's absurdity) as the person's struggle against the obsessional experience itself. He actively strives to resist the obsession, to free himself from the thought, but cannot and feels increasingly uncomfortable until the idea temporarily "runs its course" or the obsessional act has been consummated.

The illness may assume one or more of the following forms.

Obsessional ideas. Thoughts that repetitively intrude into consciousness (words, phrases, rhymes), interfering with the normal train of thought and causing distress to the person. Often the thoughts are obscene, blasphemous, or nonsensical.

Obsessional images. Vividly imagined scenes, often of a violent, sexual, or disgusting nature (images of a child being killed, cars colliding, excrement, parents having sexual intercourse) that repeatedly come to mind.

Obsessional convictions. Notions that are often based on the magical formula of thought-equals-act ("Thinking ill of my son will cause him to die"). Unlike delusions, obsessional beliefs are characterized by ambivalence: the person believes and simultaneously does not believe. As a famous psychiatrist, Karl Jaspers, expressed it, there is a "constant struggle between a consciousness of validity and non-validity. Both push this way and that, but neither can gain the upper hand."

Obsessional rumination. Prolonged, inconclusive thinking about a subject to the exclusion of other interests. The subject is often religious or metaphysical—why and wherefore questions that are as unanswerable as they are impossible to resolve. Indecisiveness in ordinary matters is very common ("Which necktie should I wear?"). Doubt may lead to extremes in caution both irksome and irresistible ("Did I turn off the gas?" "Lock the door?" "Write the correct address?"). The patient checks and rechecks, stopping only when exhausted or after checking a predetermined "magical" number of times. Obsessional doubts—*manie du doute*—may well be the most prominent feature in obsessional compulsive disorder.

As with other obsessions, ruminations are resisted. The person tries to turn his attention elsewhere, but cannot; often the more he tries, the more intrusive and distressing the thoughts become.

Obsessional impulses. Typically relating to self-injury (leaping from a window); injury to others (smothering an infant); or embarrassing behavior (shouting obscenities in church).

Obsessional fears. Often of dirt, germs, contamination; of potential weapons (razors, scissors); of being in specific situations or performing particular acts.

Obsessional rituals (compulsions). Repetitive, stereotyped acts of counting, touching, arranging objects, moving in specific ways, washing, tasting, looking. Compulsions are inseparable from the obsessions from which they arise. A compulsion is an obsession expressed as action.

Counting rituals are especially common. The person feels compelled to count letters or words or the squares in a tiled floor, or to perform arithmetical operations. Certain numbers, or their multiples, may have special significance (he "must" lay down his pencil three times or step on every fifth crack in the pavement). Other rituals concern the performance of excretory functions and other everyday acts such as preparing to go to bed. Also common are rituals involving extremes of cleanliness (handwashing compulsions, relentless emptying of ashtrays) and complicated routines assuring orderliness and punctuality. Women apparently have a higher incidence of contamination phobia and of compulsive cleaning behavior than do men.

A 49-year-old housewife is tortured by a fear of germs—the same bizarre phobia that plagued the late billionaire Howard Hughes. In a fanatic pursuit of cleanliness she uses up more than 225 bars of soap on herself every month, wears rubber gloves even to switch on a light—and makes her husband sleep alone so she won't be contaminated by him.

Every month, Mrs. X goes through 400 pairs of surgical gloves, 4,000 plastic bags—which she wears in multiple numbers over the gloves—and 360 rolls of paper towels.

She goes through dozens of boxes of laundry detergent every month because she washes her clothes six or seven times before wearing them. "And I can't bear to walk on the floors outside my bedroom. I spread newspapers ahead of me as I walk through the house. But I can't stand leaving them lying on the floor—so I leave a room by walking backward and picking up the papers in my gloved hands."

(Obsessive compulsive disorder may be utterly devastating.)

Four kinds of rituals occur most frequently: counting, checking, cleaning, and avoidance rituals. Avoidance rituals are similar to those seen in phobic disorders. An extreme example is the patient who avoids anything colored brown. Inability to approach brown objects greatly limits his activities.

Other rituals that occur less often consist of "slowing," striving for completeness, and extreme meticulousness. With slowing, such simple tasks as buttoning a shirt or tying a shoelace might take up to 15 minutes. Striving for completeness may be seen in dressing also. Asked why he spends so much time with a single button, the patient might reply that he was trying to prove to himself that he had "buttoned the button properly."

A common form of pathological meticulousness is a concern that objects be arranged in a special way. Pencils, for example, may have to be arranged so that the points are directed away from the patient. Students may spend so much time in arranging pencils, pens, erasers, etc. that they cannot do their work.

Rituals, ridiculous as they may seem to the patient, are accompanied by a profound dread and apprehension that assure their performance, since they alone give relief. "I'll explode if I don't do it," a patient may say. Occasionally a patient believes that failure to perform a given ritual will result in harm to himself or others, but often the ritual is as inexplicable to the patient as it is to the observer.

Obsessional symptoms are often accompanied by depressed mood. This may lead to an erroneous diagnosis of depression, since the mood element at the time of examination may overshadow the obsessional content.

Obsessional symptoms rarely occur singly. As with most psychiatric illnesses, obsessive compulsive disorder presents a cluster of symptoms which, individually, are variable and inconstant over time but as a group maintain characteristics unique to the illness. Thus a patient may now have one set of obsessional impulses and rituals, later another set, but the symptoms remain predominantly obsessional.

The illness may begin gradually or suddenly, usually for no apparent reason.

Obsessionals with mild symptoms requiring only outpatient therapy have a rather good prognosis; as many as 80% are recovered or improved five years after diagnosis. Hospitalized cases do less well. One-third or fewer are improved several years after discharge, and between 5 and 10% experience progressive social incapacity.

Despite the frequency with which suicide may figure in obsessional thinking, obsessionals rarely commit suicide. Obsessional patients sometimes fear they will injure someone by an impulsive act. They fear they will lose control and embarrass themselves in some manner. They worry about being addicted to drugs prescribed by their physician. These fears are generally unwarranted. There is no evidence that obsessive compulsive disorder predisposes to homicide, criminal behavior, alcoholism, or drug addiction.

Finally, obsessionals may fear they will "lose" their minds, become totally disabled, need chronic hospitalization. None of these events is common.

Obsessive compulsive disorder must be distinguished from "obsessional personality." No investigator has followed a group of clearly defined obsessional personalities over enough time to determine their fate; hence the label has no predictive value and is not in this sense a diagnosis. The individual with an obsessional personality is punctual, orderly, scrupulous, meticulous, and dependable. He is also rigid, stubborn, pedantic, and something of a bore. He has trouble making up his mind, but, once made up, is single-minded and obstinate. Many individuals with obsessive compulsive disorder have obsessional personalities antedating the illness. How often obsessional personalities co-exist with obsessive compulsive disorder is not known. Some authorities believe obsessional personality is more often associated with middle-life depressions.

Phobia-like avoidances are common in obsessive compulsive disorder. The avoidances, however, usually have a ritualistic, almost ornate quality that obviously goes far beyond simple avoidance.

Treatment

Studies on obsessional disorder justify a certain measure of optimism about its natural course. Spontaneous improvement often occurs, and the patient can be informed of this. He can be reassured that his impulses to commit injury or socially embarrassing acts almost certainly will not be carried out, and that he will not—as he often fears—lose his mind. If he needs to be hospitalized, he can be assured that the hospitalization is unlikely to be a long one.

Obsessionals are rarely helped by psychotherapy alone. "Insight" therapy may even be contraindicated. "A searching, interpretive, in-depth approach," wrote one therapist, "in many instances facilitates an introspective obsessive stance." Pessimism about psychotherapy for obsessional disorder is reflected by the scarcity of publications on the subject. Of nearly 200 articles pertaining to therapies for obsessional disorder, only 16% are studies of psychotherapy. By far the largest number are studies of behavioral treatments.

Behavior therapy derives from learning theory. According to learning theory, obsessional thoughts are conditioned responses to anxiety-provoking stimuli. Compulsions are established when the individual discovers that the compulsive act reduces the anxiety attendant on obsessional thought. The reduction in anxiety reinforces the compulsive act.

The techniques of behavior therapy are manifold but their application to obsessional disorder usually involves a single principle; the patient must be exposed to the fear-inducing stimulus (see Chapter 18). After repeated exposure, the fear disappears because, at bottom, it is unfounded. As fear subsides, so do the obsessions and compulsions. The various techniques for achieving this goal bear such names as desensitization, flooding, implosion, paradoxical intention, operant shaping, and cognitive rehearsal, all discussed in Chapter 18.

More than a dozen studies indicate that behavior therapy produces relief from compulsive rituals in many patients; the relief is maintained for at least two to three years following treatment. There is less evidence that obsessions are relieved by behavior therapy. The type of behavior therapy producing the best results is called "in vivo exposure." "In vivo" means the patients repeatedly confront the actual source of the anxiety that evokes the ritual. For example, compulsive handwashers might be asked to touch the sole of a dirty shoe and otherwise force themselves to handle dirty objects. The compulsive tidier and checker of windows might be asked to leave his or her possessions deliberately untidy and refrain from checking rituals. In many patients, the anxiety-evoking stimulus eventually will be tolerated without evoking rituals.

Exposure therapy takes up to thirty sessions in the therapist's office, plus self-exposure homework, sometimes with relatives cooperating as exposure cotherapists. Some patients are too apprehensive to submit to such an ordeal. Others are too depressed. But the evidence supporting the efficacy of exposure therapy for compulsive rituals is impressive.

Many drugs also have been used in the treatment of obsessional disorder. They include antipsychotic drugs, antidepressant drugs, and LSD.

Only one drug, chlorimipramine, showed promise as a specific antiobsessional agent. Unfortunately, the drug was recently withdrawn from the market because of potentially serious side effects.

Antianxiety drugs relieve anxiety accompanying obsessions, and antidepressants relieve depression seen with the disorder, but neither group seems to reduce the obsessional thinking.

20

Post-Traumatic Stress Disorder

> I was hung in a carriage that did not go over a bridge but which caught on one side and hung suspended over the ruined parapet. Two or three hours afterwards, among the dead and dying, surrounded by terrific sights, rendered my hand unsteady. Some time after the accident I am not quite right within, an effect of the railway shaking. . . . After the experience the shaking tells more and more, instead of, as one might expect, less and less. I am curiously weak—weak as if I were recovering from a long illness. . . . I write a half-a-dozen notes and then turn faint and sick. Riding on a train yesterday I felt more shaken than I have ever since the accident. I cannot bear railway travelling yet. I have a perfect conviction that the carriage is down on one side . . . which comes upon me and is inexpressibly distressing.
>
> Charles Dickens
> Letter to a friend

Historical Background

The idea that traumatic experiences produce a specific cluster of psychological symptoms has been debated for more than a century. Great disasters often bring about a resuscitation of the debate, usually followed by declining interest until the next disaster.

The most recent calamity reviving the debate was the Vietnam War. Several years after the war ended the Veterans Administration decided that many Vietnam veterans were suffering from something called a post-traumatic stress disorder (PTSD). The term was introduced into the official nomenclature and the VA announced that PTSD was compensable.

It was recognized that combat was not the only source of PTSD. Any psychologically traumatic event outside the range of usual human experience has the potential to produce PTSD. This includes natural disasters (floods, earthquakes), man-made disasters (car accidents, airplane crashes, fires) and experiences associated with war (bombing, torture, death camps). Some traumatic events frequently produce the disorder (torture) and others produce it only occasionally (car accidents). Rape or assault can also produce PTSD. It less often follows such common experiences as bereavement, chronic illness, business losses, or marital conflict.

PTSD was first studied in the nineteenth century following a rash of railway disasters. A man named John Erichsen wrote the first book about PTSD after studying victims of railway accidents. The symptoms were more or less as they are described today: anxiety, recurrent dreams and recollections of the event, sleep disturbance, reduced involvement with the external world. The cause, he said, was physical: a "spinal concussion," later known as Erichsen's disease. The book led to the railways losing substantial amounts of money to legal claims. A London surgeon challenged Erichsen's theory and said the cause of PTSD was purely mental and therefore *not* compensable. As evidence, he cited the frequent delay between the accident and the onset of symptoms.

Thus, three themes were introduced in the mid-nineteenth century that continue to be heard to the present day. Is the cause of PTSD physical or psychological? Should the victims be compensated? What is the explanation for the frequent delay between trauma and symptoms?

In the American Civil War PTSD was called "nostalgia," a "mild type of insanity caused by disappointment and a continuous longing for home." Five thousand cases of nostalgia required hospitalization and 58 victims died. Then the military changed its policy. Victims were jailed rather than hospitalized. The incidence of nostalgia fell to zero. Punishment as a "cure" for PTSD is another recurrent theme in the PTSD literature.

In World War I, PTSD was called shell shock. At first the cause was ascribed to neurological damage from bursting artillery shells. Then it was noted that the same symptoms occurred in soldiers not exposed to cannon fire. Thus, two types of shell shock came into existence: the legitimate type from neurological damage, and the "cowardly" type handled as a disciplinary matter.

From World War I and II came three principles for treating shell shock (PTSD), later used with striking success in the Korean and Vietnam Wars. These were *immediacy* (the soldier was treated as soon as possible), *prox-*

imity (as close to the scene of battle as possible), and *expectancy* (the soldier was expected to return to duty as soon as possible).

The Germans during World War I outlawed shell shock. Anyone presenting with shell shock would be shot. Henceforth there were no reported cases of shell shock in the German army.

Also during the World War I period, Freud introduced the concept of a "repetition compulsion" explaining many of the features of PTSD. Repetitive dreams of the traumatic event were interpreted as an attempt to "work through the psychic fixation." Earlier, Freud had commented on the frequent delay between trauma and symptoms. He also discussed predisposing factors in making some people more susceptible to PTSD than others—another recurrent theme.

Throughout World War I, disciplinary treatment vied with medical treatment as a means of managing shell shock. In France and England, shell shock manifested by a "functional" paralysis was treated by shocking the affected limb with electricity until the patient was "cured." According to Louis Leland, a Canadian psychiatrist, the longest case on record took only 3 hours to cure. He provided a case history that captures the flavor of this interesting approach:

> A 24-year-old private had been totally mute for nine months. He had taken part in the Battle of Marne, where he had collapsed with the heat and woken up mute. His condition was particularly stubborn, having resisted many therapies.
>
> Leland did the following: He strapped him down in a chair and applied a strong electric current to his neck and throat. Lighted cigarette ends were applied to the tip of his tongue and hot plates were placed in the back of his mouth. Then he took the private into a dark room, locked the doors and informed the patient that he could not leave until he acted normally.
>
> Electricity was applied for one hour, at which time the patient could say "ahh." At the end of two hours the patient tried to get out of the room but could not. The private signaled that he could tolerate no more electricity. At that point Leland raised the voltage and continued until the patient could speak in a whisper.
>
> Leland wasn't satisfied. He told the patient, "You have not recovered yet. Your laugh is most offensive to me. I dislike it very much. You must be more rational." He left the room for five minutes and returned to find the patient sober, rational, and *ready to return to duty.*

PTSD was treated more humanely during World War II, at least sometimes. Soldiers who developed the condition were told they were victims of an unhappy childhood. This awareness of predispositional factors led to rejection of individuals believed susceptible to war neurosis (PTSD). Early in the war, three out of every five potential draftees were rejected for psychiatric reasons. Despite this, psychiatric casualties in World War II were 300 times higher than in World War I.

General Patton preferred the term "combat exhaustion" to "war neurosis," holding that neurosis was a "problem of the will." He believed war neurosis should be punishable by death, an approach pioneered by the Russians with outstanding success. Eisenhower wouldn't go along.

Comparing the two World Wars illustrates the fact that symptoms of PTSD are not immutable. In World War I trembling was the primary manifestation; gastrointestinal complaints were common in World War II.

In the Korean War, combat exhaustion became known as combat fatigue, which sounded more benign. The immediacy-proximity-expectancy approach resulted in a 6% psychiatric-evacuation rate compared to 23% in World War II.

In the Vietnam War the evacuation rate was even lower. One explanation: A maximum period of 12 months of combat gave the combatant something to look forward to. In previous wars, the only way to escape combat was to get killed, wounded, or develop a combat neurosis.

THE VIETNAM WAR was an unusual war. There were no fixed battle lines. The enemy didn't wear a uniform. It was the first "air-conditioned war" in which troops alternated brief periods of intense combat with periods of near luxury in Saigon. The integrity of the fighting unit was often weak. Soldiers arrived individually and rotated home individually. Finally, the Vietnam War was profoundly unpopular. This meant the Vietnam soldiers returned home not as heroes but sometimes as near pariahs. This may account for the frequent delay in the occurence of PTSD and a VA policy toward PTSD that has received a mixed response from VA psychiatrists.

Prisoners-of-war and concentration camps have also produced many cases of PTSD. Often the symptoms developed after a period of months or years. One study found that children in concentration camps developed character traits of bitterness and dissatisfaction toward whatever was done for them; survivors who were young adults during internment later experienced chronic anxiety; older people later became depressed. Again, this illustrates the variability in PTSD symptoms—a variability that has made some people question whether the disorder even exists.

War has been dealt with at such length here because most of what is known about PTSD has come from studies of war victims. Civilian and natural disasters have not been entirely neglected, however. From studies of the Coconut Grove fire in 1941, the Buffalo Creek Dam disaster of 1972, the Mississippi Valley Flood of 1973, the Chowchilla Kidnapping of 1976, and the Hyatt Regency Hotel disaster of 1981 have emerged some fairly consistent observations.

1. Claims for damages for PTSD are directly related to the incidence of PTSD. Lawyers, perhaps more than physicians, contribute to this increased incidence.

2. Survivors often feel guilty because they escaped while others perished.

3. Delayed symptoms are even more distressing than the acute reaction to trauma. Half or more of victims continue reporting symptoms a year after the disaster. However, few require hospitalization. Many receive help from a general practitioner. Typically, the victims do not regard themselves as psychiatrically ill.

4. Omens are sought: looking back for ways that one could have avoided the calamity. Many people who go through a disaster believe they had been given a sign before it happened.

Primitive countries have elaborate rituals for expunging the psychological after-effects of trauma. Danger from the ghost of the slain is a frequent theme. Young braves of Indian tribes who had taken their first scalps were obliged to practice abstinence for six months. They were not to sleep with their wives and could only eat fish and hasty-pudding. The Arunta of Central Australia believe that the spirit of a slain man follows the tribe and does mischief to those who performed the slaying. The spirit takes the form of a little bird called the chichurna, which may be heard crying like a child in the distance. The warriors lie awake at night listening for the cry of the bird in which they fancy they hear the voice of their victim. As psychiatrist William Rinck has pointed out, this reminds one of the Vietnam warrior, unable to sleep, haunted by battle nightmares he experiences for nights without end.

Rinck found that similar rituals have been reported the world over:

1. The returning warrior is considered unclean and must undergo some form of ritual purification. Often this is done apart from the village in a ritual hut. In our recent past, triumphal arches in Rome and Paris recall an era when warriors were ritually honored. America had ticker-tape parades and eulogies for its returning warriors.

2. Ritual appeasement of the ghosts of the slain was ubiquitous. Perhaps this is symbolically recognized in the financial generosity toward defeated enemies that countries often display. As Rinck points out, perhaps the Vietnam soldier hasn't undergone the ritual purification that would permit him to come to peace with himself and his society.

This long historical note was believed justified because PTSD, compared to the other anxiety disorders, is unique: many people—as noted—do not believe it exists. They believe traumatic experiences produce a variety of unpleasant psychological after-effects but the symptoms lack cohesion or a stable course over time that characterizes a distinct clinical entity (disease, or disorder). Many believe that a better name for PTSD is compensation neurosis, a topic discussed later.

In any case, the authors of DSM-III—the official classification system of the American Psychiatric Association—disagreed. After much debate they introduced PTSD into the official nomenclature in 1980 with the following features.

Diagnostic Criteria

A. Existence of a recognizable stressor that would evoke significant symptoms of distress in almost anyone.

B. Reexperiencing of the trauma as evidenced by at least one of the following:

1. Recurrent and intrusive recollections of the event
2. Recurrent dreams of the event
3. Sudden acting or feeling as if the traumatic event were reoccurring, because of an association with an environmental or ideational stimulus

C. Numbing of responsiveness to or reduced involvement with the external world, beginning sometime after the trauma, as shown by at least one of the following:

1. Markedly diminished interest in one or more significant activities
2. Feeling of detachment or estrangement from others
3. Constricted affect (show of emotion)

D. At least two of the following symptoms that were not present before the trauma:

1. Hyperalertness or exaggerated startle response
2. Sleep disturbance

3. Guilt about surviving when others have not, or about behavior required for survival
4. Memory impairment or trouble concentrating
5. Avoidance of activities that arouse recollection of the traumatic event
6. Intensification of symptoms by exposure to events that symbolize or resemble the traumatic event

Subtypes of PTSD also were described. Among them was *Post-Traumatic Stress Disorder, Delayed*, with onset of symptoms at least six months after the trauma.

Elaboration

PTSD has three major symptoms: (1) persistent recollections of the event; (2) a reduction in responsiveness to everyday events, sometimes called "psychic numbing" or "emotional anesthesia"; (3) an exaggerated startle response.

As with Charles Dickens, who survived a train wreck, PTSD victims experience vivid recollections of the traumatic experiences whenever they are exposed to situations or activities that resemble the original trauma, such as hot, humid weather or tall grass for veterans who fought in Southeast Asia or the South Pacific. They complain of feeling detached or estranged from other people. They lose interest in activities that previously interested them and experience emotional blunting, so they feel less tender or intimate with wives and lovers. The exaggerated startle response is an example of a general hyperalertness. They have trouble falling asleep. They have recurrent nightmares, particularly about the traumatic event, and they have trouble concentrating or completing tasks. They also describe guilt feelings about surviving when many did not, or about the things they had to do in order to survive.

Anxiety is common, although the physical symptoms of anxiety, such as palpitations and breathlessness, are not. Feelings of depression are also common, often manifested by irritability and explosions of aggressive behavior without provocation. Sometimes the victims behave impulsively, disappearing from work or home, taking long trips, making changes in residence for no obvious reason. Survivors of death camps have particular difficulty with memory and concentrating.

The symptoms may begin immediately or soon after the trauma. It is not unusual, however, for the symptoms to emerge after a latency period of months or years following the trauma.

Impairment may be mild or devastating. Phobic avoidance of situations or activities resembling or symbolizing the original trauma may result in occupational or recreational impairment. "Psychic numbing" may interfere with marriage or family life. Suicide and suicide attempts have been reported.

It is important to emphasize that not everyone who undergoes a severe traumatic experience expresses symptoms of PTSD. The majority of Vietnam veterans who experienced combat do *not* believe the war had a bad effect on their personal development; many felt they had been strengthened by the war and made tough and self-reliant.

There is evidence, however, that the more severe the combat experience the more likely that post-traumatic symptoms will occur. Psychiatrist Nancy Andreasen compares the role of the trauma in PTSD to the role of force in producing a broken leg. She writes: "It is normal for a leg to break if enough force is applied, although a broken leg is a pathological condition. Individual legs vary, however, in the amount of force required to produce a break, the amount of time required for healing, and the degree of residual pathology that may remain. In most persons experiencing post-traumatic stress disorder, the stressor is a necessary but not sufficient cause, because even the most severe stressors do not produce a post-traumatic stress disorder in all persons experiencing the stressor. A variety of psychological, physical, genetic and social factors may also contribute to the disorder."

Pre-existing psychiatric disability may increase the impact of particular stressors. Studies have shown that persons with a prior history of psychiatric treatment have a greater probability of developing a post-traumatic syndrome. A family history of psychiatric illness also predisposes a person to develop PTSD. Individuals with PTSD have a high rate of alcoholism but studies show that almost all the affected individuals were heavy drinkers before experiencing the traumatic experience.

Most cases of PTSD are acute, with the symptoms persisting for less than six months. Usually the symptoms resolve spontaneously without psychiatric treatment.

Chronic PTSD is less common. The more chronic the symptoms, the worse the prognosis. The possibility of receiving compensation also worsens the prognosis. It is a common observation that the patient's symptoms and his perception of his illness are often heavily influenced by what previous physicians or lawyers have told him. This particularly applies to the subcategory of post-traumatic stress disorder called PTSD, *delayed type.*

Compensation Neurosis

This is an old term for a condition in which symptoms of physical or mental disorder seem to be influenced by personal gain from having the disorder. For many psychiatrists PTSD often represents a form of compensation neurosis. The history of PTSD provides the background for this conviction.

In 1980 the Veterans Administration announced that post-traumatic stress disorder, delayed type, was a compensable disorder. This meant that for the first time since World War I, the Department of Veterans Benefits could consider disorders to be service-connected when the symptoms appeared long after military discharge.

Many veterans responded by filing claims based on their belief that they suffered from post-traumatic stress disorder, delayed type, related to traumatic war experiences. The symptoms of the disorder were well publicized in the media and in brochures distributed by national service organizations. Rarely before had so many claimants presented themselves to psychiatric examiners having read printed symptom checklists describing the diagnostic features of the disorder for which they sought compensation. Some psychiatrists were offended by the "mechanical litany of complaints recited by a well-read claimant" (as one psychiatrist put it.)

Very few psychiatrists question the legitimacy of acute and even chronic post-traumatic stress disorders. It is the delayed type that produces controversy. It is important to note that a delay between trauma and symptoms was noted long before the Vietnam War. At least a half dozen studies after the second World War comment on the delay. As noted earlier, Freud observed the delay following civilian disasters at the turn of the century.

Nevertheless, only American psychiatrists have introduced the concept of delay into the *definition* of a post-traumatic stress disorder. There is no equivalent category in *The International Classification of Diseases* used in other countries. The closest equivalent is "acute reaction to stress" displayed by "very transient disorders which occur . . . in response to exceptional physical or mental stress . . . which usually subside within hours or days." This definition is almost identical to the post-traumatic stress disorders described in earlier versions of the American classification. Apparently, certain unique features of the Vietnam War, commented on earlier, were required for the concept of a delayed reaction to be included in the definition, plus a more elaborate description.

Delay, in the older literature, did not always occur in the context of compensation, but in the case of Vietnam veterans, this connection has become inevitable. How do psychiatrists determine whether the delayed form of PTSD is involuntary and thus compensable or represents a form of malingering?

Malingering is a word psychiatrists rarely use and with good reason. It is almost always impossible to know truly whether a person's actions are entirely voluntary or involuntary. Hypnosis, drugs, and psychological tests have all failed to help the clinician judge whether a mental illness is being faked. A lack of willingness to suspect malingering is a major problem. It seems to violate the doctor-patient relationship. Doctors are as reluctant as the next person to call somebody a liar—and they may get sued if they do.

DSM-III defines malingering as "the voluntary production of false or grossly exaggerated symptoms in pursuit of a goal that is obviously recognizable with an understanding of the individual's circumstances." Thus malingering requires a deceitful state of mind. No other syndrome is as easy to define but as difficult to diagnose.

Mental illness pretenders have a variety of goals: to escape from punishment via the insanity defense, to gain compensation for mental disability, to obtain drugs or a transfer to a psychiatric unit while a prisoner, and even to seek shelter from the cold.

Because the symptoms of PTSD are almost entirely subjective, a resourceful malingerer can quite convincingly report the "right" symptoms. Kathleen Quinn, a psychiatrist at Case Western Reserve, provides some advice for verifying the patient's symptoms. First, ask the spouse about the reported nightmares and the trauma experience itself. Other clues to malingering: "claimed inability to accept any kind of work, yet tenacious pursuit of compensation benefits and continuing active involvement in recreational activities; refusal to cooperate in recommended diagnostic or therapeutic procedures unless they are required for receiving benefits; unwillingness to make definite statements about returning to work or other personal expectations; and exclusive complimentary self-depiction despite a history of spotty employment and drifting."

By 1984, the VA had paid compensation to more than 7,000 Vietnam veterans with a diagnosis of post-traumatic stress disorder. This is about 1% of the 700,000 estimated by the VA to have the disorder. Most psychiatrists would probably view 700,000 as an overestimate, but of course there is no way to prove it. Most psychiatrists believe that post-traumatic stress disorder, delayed type, is a legitimate diagnosis and, in most cases, not influenced by compensation. The evidence for this resides chiefly in earlier studies showing that delay was a frequent occurrence in stress disorders where compensation was not a factor.

Treatment

Many disaster victims refuse psychiatric treatment. They feel their symptoms are understandable in view of the trauma they underwent and assume the symptoms will pass in time, which they usually do.

Different disasters have produced different combinations of symptoms. The rape victim, the Holocaust victim, the prisoner of war, and the Vietnam veteran all have particular types of problems that do not necessarily overlap. The post-traumatic stress disorder described here was a product of the Vietnam War.

The first Vietnam veterans treated for PTSD were involved in war crimes. A Yale psychiatrist, Robert Jay Lifton, had experience with trauma victims, having studied survivors of Hiroshima. He began working with Vietnam veterans and found that "rap groups" (a form of group therapy) seemed to help the men. He attributed the success to the fact the participants had shared a particular historical experience; each functioned as a therapist; and all assumed the responsibility for the shape and direction of the enterprise. The participants were particularly bitter toward chaplains and psychiatrists. They referred to the hypocracy of chaplains who blessed whatever they did, including atrocities. "Help" by a psychiatrist was defined as being returned to duty. Lifton observed: "It was one thing to be ordered by command into a situation that came to be perceived as both absurd and evil, but it was quite another to have that process rationalized and justified by the ultimate authorities of the spirit and mind—that is, by chaplains and psychiatrists."

Whether rap sessions were actually effective was never studied but nevertheless they became the central mode of therapy for Vietnam veterans with PTSD. By 1984 about 150 Vet Centers for treating possible PTSD had opened across the United States. Based on directives from the VA Central Office, here are some of the adjustments required of the traditional mental hygiene clinic in dealing with possible PTSD victims: Service was to be rapid; there could be no waiting list. No records were to be kept. Without records there could be no statistics and thus no way of determining how many veterans had PTSD. However, it became clear that nearly half of the patients had problems with alcohol and most had job problems. Treatment of the alcohol problems and job rehabilitation became an important part of the program.

Treatment of disaster victims in general is symptomatic. Sleeping medication may prevent nightmares. Antianxiety drugs may reduce the hyperalertness and exaggerated startle response. Antidepressant drugs are helpful both for phobic symptoms and feelings of depression. Behavioral techniques, such as systematic desensitization and relaxation therapy (see Chapter 18), are useful in preventing phobic avoidances. Occupational counseling is often needed. Psychotherapy is usually brief and supportive.

Many patients who develop PTSD had pre-existing psychiatric illnesses. A psychiatric examination is essential to identify illnesses that existed before the traumatic event and which may be treatable.

Notes and References

CHAPTER 1

P. 3

The quotations from Rollo May, unless noted otherwise, appeared in *The Meaning of Anxiety*, published by W. W. Norton in 1950. It was revised in 1977 and appeared in paperback (Pocket Books).

PP. 3–6

The distinction I make between fear and anxiety is not one always found in the dictionary. Kierkegaard and Freud defined the words as I do, and most psychiatrists, when they think about it, probably do likewise. The distinction applies only to humans. "Fear," when applied to animal behavior, refers to a defensive response that resembles human responses associated with the subjective experience of fear. When a pearl fish flees from danger into the anus of a sea cucumber, this is called a fear response on the grounds that, in circumstances where this was the only haven available, a human would probably do the same.

Fear assumes a number of guises that have different names. Not all authorities agree that they represent different forms of fear, but enough do to justify making the distinctions.

Guilt is a type of fear: the fear that you have done something wrong (you may not know what) and may suffer from it. What is "wrong" is culture-bound. Huckleberry Finn felt guilty because he helped Nigger Jim escape slavery—and now would feel guilty if he hadn't. Guilt, in our time, is perhaps the fear most of us experience most often. In *How to be a Jewish mother* by Dan Greenberg (Random House, 1964), mothers are advised to make their sons feel guilty: "If you don't know what he's done to make you suffer, *he* will."

Shame is the fear of being seen naked—perhaps not *really* naked (although even stripteasers have unpleasant dreams of being unclothed in public) but naked in the sense that people see us as we are: stupid, inept, unattractive, worthless (and most of us fear we are one or all of these from time to time).

Jealousy and *envy* are fears. The first is a fear of losing what you have (eroticized in the case of spouses and lovers) and the second is a fear of somebody else being treated better than you are: a favorite child.

These are emotions and what is an emotion? It is perhaps the hardest word of all to define. Seen from outside, emotion suggests a person in a special state: he is "e-moved," "moved out of himself," likely to act for a time in a more or less unusual way.

From the inside, emotion is a feeling that prompts action of some kind, as opposed to a *thought*, which judges whether the action is possible, justified, or in one's best interest. Feelings and thoughts are often at odds: the thinker cannot decide what is best and the feeling is mixed, contradictory, a jumble of fear, anger, and other ingredients of an emotional goulash.

These definitions are not in the dictionary either. Definitions, at bottom, are arbitrary. "When I use a word," Humpty Dumpty told Alice, "it means just what I choose it to mean—neither more nor less." The above words mean, more or less, what I choose them to mean, and let those who argue, argue with Humpty Dumpty.

For the etymologist, three words used repeatedly in the book—fear, anxiety, and panic—have interesting origins:

Fear comes from the Old English, *faer*, meaning sudden danger. It refers to fright where fright is justified. The danger is concrete, real, knowable. The fear is appropriate, and sometimes useful, if one is to escape harm.

Anxiety comes from the Latin, *anxius*, meaning a tight feeling in the chest. It refers to a fear of uncertain origin. The person does not know why he is afraid—or his fear seems disproportionate to the danger. "I just feel anxious," he says, often with a tight feeling in the chest.

Panic refers to extreme fear. The term comes from Pan, the Greek rural deity. Pan is sometimes a friendly god, looking over sheep and shepherds, and a music-lover. But Pan also sometimes scares the hell out of people. In short, Pan is a mixed blessing, and panic, too. Panic may head you straight for the exit in case of fire—but sometimes the wrong exit.

The most detailed and closely reasoned arguments for distinguishing fear from anxiety are found in the works of Kurt Goldstein (e.g., *Human Nature in the Light of Psychopathology*, Harvard University Press, 1938). Goldstein writes:

> In fear, there is an appropriate defense reaction, a bodily expression of tension and of extreme attention to a certain part of the environment. In anxiety, we find meaningless frenzy, with rigid or distorted expression, accompanied by withdrawal from the world . . . in the light of which the world appears irrelevant. . . . Fear sharpens the senses, whereas anxiety paralyzes the senses.

Anxiety, he writes, tends to confuse not only one's awareness of one's self but also confuses one's perception of the objective situation. This is understandable, Goldstein says, for to be conscious of one's self is "only a correlate of being conscious of objects. *The awareness of the relationship between the self and the*

NOTES AND REFERENCES   193

world is precisely what breaks down in anxiety [emphasis added]; hence it is not at all illogical that anxiety should appear as an objectless phenomenon."

P. 5

Freud's first theory of anxiety is best described in his 1915 book, *Introductory Lectures on Psychoanalysis* (Standard Edition, J. Strachey (ed)., Hogarth Press, 1901). He revised the theory in 1926 in a book called *The Problem of Anxiety* (see Chapter 6).

CHAPTER 2

P. 7

Kierkegaard's *The Concept of Dread* was originally published in Danish in 1844. In 1944 a translation by Walter Lowrie was published by the Princeton University Press.

P. 7

The quotes from Hippocrates are from section LXXXII of the fifth volume of *Epidemics*. The quotation by Burton is from the 11th edition of *The Anatomy of Melancholy*, published in 1813.

P. 8

The passage from Pascal appeared in his *Pensees*, edited and translated by G. B. Rawlings (Peter Pauper Press, 1946.)

P. 9

The quotations from Kierkegaard are from *The Concept of Dread*.

P. 10

The remarks by Walter Kaufmann are from his *Introduction to Existentialism: from Dostoevsky to Sartre* (Meridian, 1956).

P. 11

The passage about the French Revolution came from *The Days of the French Revolution* by Christopher Hibbert (William Morrow, 1980).

P. 12

Existential despair is not limited to plays by Sartre and Samuel Beckett. Detective story writers also show the influence. One of the better of the genre, Nicholas Freeling, describes a typical day's life of one of his detective heroes (*The Book of the North Wind*, Viking, 1983).

Each new day you start from scratch. Each day you know nothing. Learn something. You got up in the morning and made the coffee: it was a new page in the notebook, white and clean. You began neatly, writing legibly. Through the day the page got steadily dirtier, stained and scribbled: the phone numbers in margins, the balloons containing 'Thinks,' and the balloons surrounding 'He says'—generally meaning don't-believe-a-word-of-it. Mysterious drawings made by the unconscious while Holding the Line. Dirt accumulated. Bits got torn off, bits crossed off: bits of lives, and much of that his own, nibbled away. The end of the day was dogeared and greasy: enough was enough.

P. 12

Whether there is a decline in religious faith is questionable. In the 1920s and 1930s Robert and Helen Lynd did a classic study of a town they called Middletown (really Muncie, Indiana), reporting such a decline. The study was repeated in the 1980s by Caplow et al. The recent researchers found that conventional forms of piety were flourishing; indeed, levels of religious observance and practice appeared to be higher now than two generations ago. They concluded that claims of religious demise in the modern world had been exaggerated (*All Faithful People* by Caplow et al., University of Minnesota Press, 1983).

When Rollo May wrote his book on anxiety in 1950, he cited the Lynds' study as evidence that a deterioration of traditional values produced a corresponding increase in anxiety. The Caplow study raises questions about whether May's premise (shared by so many) is valid.

P. 12

Just how *bad* the good old days used to be was described by Robert Darton in "The Meaning of Mother Goose," an article published in the February 2, 1984, issue of *The New York Review of Books*.

CHAPTER 3

P. 14

Walter B. Cannon's autobiography, *The Way of an Investigator*, was published by Hafner Publishing Company (New York) in 1945. It is delightful reading and, if you can find a copy, it would make a good gift for a medical student or young scientist.

Cannon's book, *Bodily Changes in Pain, Hunger, Fear and Rage*, was published in 1927 by Appleton-Century-Crofts (New York). His comments on hunches appeared in the autobiography. His term "extraconscious" never caught on, perhaps unfortunately.

P. 16

The Startle Pattern was published in 1939 by C. Landis and W. A. Hunt (New York: Holt, Rinehart & Co., Inc.).

P. 17

Facial Expressions: On the Expression of Emotions in Man and Animals by Charles Darwin (London: John Murray, 1982).

P. 17

Translating Yoruba: Psychiatry Around the Globe: A Transcultural View by Julian Leff (Dekker, 1981).

P. 17

The quotation by Burton is from the 11th edition of *The Anatomy of Melancholy*, published in 1813.

P. 18

Fight and flight are not the only possible responses to danger. Immobility is a third. It is often observed in animals. Rather than running away or fighting back, the animal "freezes"—standing absolutely still, rolling in a ball, sometimes pretending to be dead. The emergency responses orchestrated by the sympathetic nervous system are replaced by defensive responses controlled by the parasympathetic system. Instead of speeding up, the heart rate becomes slower and may even lead to cardiac arrest and death. This is sometimes called a "vagal" death, because the vagus nerve—a branch of the parasympethetic system—makes the heart beat slower. What is called "voodoo death" in humans may involve the same mechanism. Immobility may actually have survival value. There are many reports of predators passing by immobile animals to pursue fleeing ones. When a human becomes "paralyzed with fear," the physiology may be the same as occurs in animals "playing possum," and there are certain situations where a Ghandi-like passivity turns away wrath better than fleeing or fighting.

P. 20

The James-Lange theory is described in William James' *The Principles of Psychology* (Henry Holt & Co., 1893, Volumes I and II).

P. 20

The studies of paraplegics and quadraplegics was cited by S. Schachter in his 1964 paper, "The Interaction of Cognitive and Physiological Determinants of Emotional State," which appeared in *Psychobiological Approaches to Social Behavior*, edited by P. H. Leiderman and D. Schapiro (Stanford University Press, 1965).

PP. 20–21

The study showing fast heart rates in patients with "anxiety states" even when not anxious was conducted by B. V. White and Edwin Gildea and reported in *The Archives of Neurology and Psychiatry* 38:964–984.

P. Tyler, I. Lee, and J. Alexander reported the increased awareness of heart rate in anxious patients in *Psychological Medicine* 10:171–174.

P. 21

Paul Thomas Young, in his article on "Emotion" in the *International Encyclopedia of Social Sciences*, Volume 5 (New York, 1968) explains how subjective interpretation of a threatening situation influences specific physiological responses.

P. 22

Oxford University Press published the classical observations of Stewart Wolf and H. G. Wolff of Tom's gastric reactions in *Human Gastric Function* (1943).

CHAPTER 4

P. 23

"Some fears are innate." This is true in a general sense, but one should be wary of attributing behavior to heredity simply because it is hard to think of another explanation. Consider this: fetuses can hear perfectly well from the uterus. Loud, angry, voices *may* induce fear responses in the fetus just as they do in the infant. Conceivably, if the emergency responses of the fetus are activated repeatedly or at critical periods in development, there may be lasting effects on the nervous system. Is it possible that one child is a bully, and another timid, and a third hyperactive because of overheard conversations during fetal life? There is not a shred of evidence for this, but, yes, it is possible.

To repeat: not everything that seems inherited is really inherited. The disease kuru is a fine example. It is a disease of the nervous system that resembles multiple sclerosis and occurs only in natives of a small South Pacific island. It occurs in middle life, is almost always fatal, and runs in families. All the scientists agreed that it was an inherited disease—until, one day, someone suggested an environmental explanation. It turns out these particular natives are cannibals who eat only the brains of close relatives. If these brains contain a microorganism called a "slow virus," now believed to cause kuru, then the disease would be passed on from generation to generation through dietary habits and not genes.

P. 24

The "hawk effect" is one of the oldest observations in ethology, first described more than 100 years ago. Konrad Lorenz wrote about it in the early 1950s and the description given here comes from the book by Adam Smith called *Powers of Mind* (Random House, 1975).

The first experimental studies of the hawk effect were conducted in the late 1950s at the Regent's Park Zoo (London) by Melzak, Penick, and Beckett (*Jour-*

nal of Comparative and Physiological Psychology 42:694–698). Dr. Penick, now a Professor at Kansas University, recalls the studies as "fun," despite "the soggy British weather, the interminable tramway rides to Regent's Park Zoo, and the double-vision caused by intently observing the antics of small feathered creatures as cut-out shapes whirled above their heads." (She has the office next to mine, which is how I know these things.)

P. 24

A fear of strangers clearly has survival value. The term "child abuse" conjures up images of cruel parents, but actually children are more likely to be abused by strangers—or killed—than by a mother whose "maternal instinct" almost always, fortunately, overrides her homicidal impulses.

P. 25

The observations by Hebb are from A Textbook of Psychology (W. B. Saunders, 1964).

P. 26

The quotation from Stanley Hall appeared in the American Journal of Psychology 8:147–239.

The LSD user was Adam Smith.

P. 28

The mixed-up raven is found in King Solomon's Ring by Konrad Lorenz (Methuen, 1952).

P. 29

The Hebb quote is from A Textbook of Psychology (W. B. Saunders, 1964).

P. 31

Summation may partly explain the intense, unreasonable fears called phobias. Innate fears usually do not reach phobic intensity but may lay the groundwork for a phobia if reinforced by events in later life.

For example, a three-year-old is afraid of strangers but gets over it by five. Then, at 12, he bungles a class recitation and thereafter has a phobia about public speaking. At five, a child is frightened by dogs but not for long; every neighbor has one and she likes her own dog too much to be afraid of them. Then, at 20, a dog bites her and she develops a dog phobia.

Innate fears and summation may explain why some phobias are more common than others, but obviously this is not the whole story. People are bitten by dogs every day and most do not develop dog phobias.

In terms of arms, legs, opposing thumbs, and maybe certain fears, we are all born with similar characteristics but also with conspicuous differences. Some legs are short, some long, some straight, some bowed, etc. The same is true of temperament. Tiny babies have strikingly different temperaments, ranging from stolid to stormy. The differences cannot be attributed to differences in maternal care; obstetricians spot them at time of delivery. Our temperaments may be as molded by heredity as our hairlines.

It was true of Pavlov's dogs (Chapter 5). Some were easily conditioned and others not, and he blamed "temperament." Identical twins provide further evidence. In those rare instances when identical twins are separated at birth and raised apart in different environments, they show striking temperamental similarities in adulthood. If one is shy, the other is shy; if one is anxious, the other is anxious; and if one has phobias, the other tends to have phobias.

Does this mean that specific phobias are inherited? Probably not. There is some evidence that agoraphobics have an increased rate of various psychiatric illnesses in their families, but the evidence is mixed and not impressive.

What is more likely, perhaps, is that people vary in their innate susceptibility to anxiety. Anxious people, for example, develop a conditioned response faster than unanxious people. To the extent that conditioning plays a role in phobias, for example, anxious people may be phobic-prone.

CHAPTER 5

P. 32

According to a recent newspaper article, a San Francisco psychologist and two colleagues have determined that "the mechanics of facial muscle movement are closely tied to the autonomic nervous system, which controls heart rate, breathing, and other vital involuntary functions." Other studies have shown that people tend to mimic the facial expressions—including smiles—of those around them. Reading this, a writer for *The New Yorker* was "happy to get confirmation of a feeling I have often had; namely, that we are all mixed up in one another, and that we murmur continuously to one another in a variety of subtle languages, changing one another's blood pressure, sending heart rates up and down, affecting one another's muscle tension and breathing. In some essential way, we are always speaking the truth to one another and—on a physical level, at least—always understanding one another."

In short, we are the conditioners and we are the conditioned, in the most complicated ways, in our social living.

P. 34

The Mark Keller quote is cited in *Alcoholism: The Facts* by D. W. Goodwin (Oxford, 1981).

NOTES AND REFERENCES

P. 34

About one-trial learning: Some people attribute their phobias and other anxiety disorders to a single stressful event. This happens most often with simple phobias. A person is stuck in an elevator and therefore avoids elevators; a dog bite leads to a dog phobia; a young person learning to drive dents the family car, is scolded, and refuses to drive again.

Obviously, the traumatic event does not tell the whole story: many more people are stuck in elevators than develop elevator phobias. Most phobias, moreover, are not associated with a single frightening event—at least one the victim recalls.

Perhaps, however, the victim has forgotten the traumatic event. Perhaps it happened when he was a small child and he repressed the memory. Or, even as an adult, perhaps people actively put out of their conscious mind thoughts that are too upsetting.

Some years ago, there was a theory, popularized in novels and World War II movies, that *forgetting* a traumatic event was more important than the trauma itself in causing a neurotic disturbance such as a phobia. A soldier would see his buddy killed and repress the memory only to have it expressed symbolically as blindness, paralysis, or another form of "hysteria," or, in some cases, severe phobias. The treatment consisted of the army doctor bringing the patient's memory back to consciousness—sometimes with the help of hypnosis or sodium amytal. This dramatic cure, called "catharsis" or "abreaction," differed from the early Freudian cure based on "insight." The latter came from the theory that phobias were caused by anxiety resulting from unconscious childhood sexual conflicts; bringing the memory of these into conscious awareness was the heart of the treatment but a far more lengthy and complicated process was required than occurred on the movie screen—and sometimes in real life.

Cathartic cures are not as fashionable now, probably because they usually do not work. Today's version of shell shock (World War I) and war neurosis (World War II) is something called the "post-traumatic stress disorder" and is discussed in Chapter 20.

P. 37

Gossip is considered a minor vice, whereas, on the contrary, it is one of the nicest things we do.

The novelist Dan Davin explains why:

> Gossip, talking behind people's backs, is a necessary ingredient of our social culture, indispensable to it and at its best one of the finest instruments of our civilized living. We all know that we are talked about but . . . we are able to be unaware of it most of the time as we are, happily, unaware of death. We can gossip with enjoyment about others, serenely forgetful that they at the very same moment may be gossiping with equal enjoyment about us.

For gossip is a necessary exercise: it is the practice of our skills in social perception, the daily do-it-yourself art by which we refine and sharpen our sensibilities, subtilize our knowledge of others, check our knowledge by the competing perceptions of our cronies, remind ourselves of the relativity of all human truth and the exhilarating improbabilities of behavior.

Through gossip we acquire the sense of others without which we fail as social beings and which is the *sine qua non* of charity.

P. 38

Music is a soteria *par excellence:*

> And so I sing,
> As the boy goes by
> The burying ground,
> Because I am afraid.
> *Emily Dickinson*

CHAPTER 6

P. 41

Freud's theories of anxiety are described in *The Problem of Anxiety*, translated by H. A. Bunker and published by W. W. Norton in 1964 (originally published under the title *Inhibition, Symptom and Anxiety* by the Psychoanalytic Institute, Stamford, Conn., 1927).

P. 43

The story of Little Hans can be found in *Freud's Collected Works*, Volume 3 (Hogarth Press, 1909).

P. 44

Freud's gift for finding sexual significance in street traffic and shopping is developed by R. P. Snaith in "A Clinical Investigation of Phobias" in the *British Journal of Psychiatry* 114:673–677.

P. 45

The philosopher Karl Popper eloquently defends the idea that science can only prove that something is *untrue* and never that something is true, and that "falsifiability" is the test of a good scientific theory (see Chapter 9).

P. 45

The story of Little Albert was first described in a 1920 paper in the *Journal of Experimental Psychology* (3, 1–14) called "Conditioned Emotion Reactions."

P. 46

B. F. Skinner coined the term "operant conditioning." The real father of operant conditioning, however, was A. H. Thorndike, who formulated the "law of effect" in his Ph.D. thesis in 1898.

CHAPTER 7

P. 51

Paul Ehrlich's contributions were really to chemotherapy rather than bacteriology. He found that arsenic killed the organisms that produced sleeping sickness and syphilis but was not effective for bacteria and viruses. Nevertheless, his theory that altering a chemical's structure, even subtly, would alter its effects laid the groundwork for the great age of antibiotics (*Disease and Its Control* by R. P. Hudson, Greenwood Press, 1983).

P. 51

Leo Sternbach tells how he discovered Librium and Valium in *The Benzodiazepine Story*, published in 1978 by Roche. Further details are available in *Discoveries in Biological Psychiatry*, edited by F. J. Ayd, Jr., and B. Blackwell (Lippincott, 1970).

P. 53

The origin of serendipity is uncertain. One authority claimed the word "signified a mental state in which serenity and stupidity are blended . . . for example, the serendipity of a cow chewing its cud under a shady tree." Not so, according to Theodore Remer, who wrote a book on the subject. Remer says the word was coined by Horace Walpole in a letter to a friend written in 1754 and came from a Persian fairy tale called the "Three Princes of Serendip" (or Ceylon) whose heroes made fortunate discoveries by accident.

P. 56

Irvin M. Cohen tells how the benzodiazepines were clinically tested and how "serendipity might be said to have worked in reverse, namely, at two points in its development the drug came near to being overlooked or scrapped." That the compound was saved from such a fate was due, in part, to an energetic researcher who persisted in reinvestigating it, despite its failure in its initial clinical trials.

P. 57

Drugs rapidly absorbed into the body and rapidly eliminated—nicotine and heroin being examples—seem more addictive than drugs that become concentrated more gradually in the blood, such as phenobarbital and Librium. This raises the question of whether the newer short-acting benzodiazepines, such as Ativan and Xanax, will ultimately be more abused than the older long-acting benzodiazepines such as Librium and Valium. Experience with the drugs has been too brief to draw a definite conclusion, but, after several years of use, they appear relatively safe.

P. 58

Although benzodiazepines have a taming effect on animals, they sometimes release aggressive tendencies in humans. Curiously, the drugs have an antiaggressive effect in mice but not in rats. This illustrates the difficulty in extrapolating drug effects from species to species. Benzodiazepines, in any case, are not the first drugs to both tranquilize and release aggression; alcohol comes to mind. To the extent that benzodiazepines release aggression in humans, the effect is slight and probably of no social or clinical significance. ("The Effects of Benzodiazepines have a mild abuse potential compared to most other mood-Press, 1973).

PP. 58–59

How much Librium, Valium, or other benzodiazepines are required to produce death is not known, but the amounts are huge. One investigator, only half facetiously, said the only way a benzodiazepine would kill anyone would be to fill up two or three trucks with pills and dump them on the patient, causing suffocation.

P. 59

"A very small percentage of patients who take benzodiazepines abuse them." To evaluate the extent to which they are abused, the following articles are recommended: "Dependence on Benzodiazepines" by Lader in *The Journal of Clinical Psychiatry*, April 1983; "The Nature and Extent of Benzodiazepine Abuse: An Overview of Recent Literature" by Dietch in *Hospital and Community Psychiatry*, December 1983; "Current Status of Benzodiazepines" by Greenblatt, Shader, and Abernathy in *The New England Journal of Medicine*, August 11, 1983, and August 18, 1983; and "Long-term Diazepam Therapy and Clinical Outcome" by Rickels, et al. in *The Journal of the American Medical Association*, August 12, 1983. Reading these articles, one can only conclude that benzodiazepines have a mild abuse potential compared to most other mood-altering drugs.

This was further confirmed in a study published in *The Journal of the American Geriatric Society* in August, 1984, in which a review of benzodiazepine use in a group of elderly patients indicated that prolonged use, at low doses, appeared not to be harmful. The study population consisted of 93 patients aged

NOTES AND REFERENCES   203

55 years or more who had used benzodiazepines over long periods of time. They reported the medication calmed them, relieved tension, and helped them sleep, while lightheadedness and sedation were the most commonly reported side effects. None reported increasing the dose of the drug they used. The study concluded that "addiction to [benzodiazepines] is exceedingly rare."

P. 60

Whether heroin got its name because it offered a "heroic" cure for morphine addiction is not certain. The word comes from the Greek *heros* and the word may have been chosen because of the feeling of omnipotence experienced while taking the drug.

P. 61

Aldous Huxley supports the position of Nathan Eddy, and elaborates:

> That humanity at large will ever be able to dispense with Artificial Paradises seems very unlikely. Most men and women lead lives at the worst so painful, at the best so monotonous, poor and limited that the urge to escape, the longing to transcend themselves if only for a few moments, is and has always been one of the principal appetites of the soul. Art and religion, carnivals and saturnalia, dancing and listening to oratory—all these have served, in H.G. Wells' phrase, as "doors in the wall." And for private, everyday use, there have always been chemical intoxicants. All the vegetable sedatives and narcotics, all the euphorics that grow on trees, the hallucinogens that ripen in berries or can be squeezed from roots—all, without exception, have been known and systematically used by human beings from time immemorial. And to these natural modifiers of consciousness modern science has added its quota of synthetics—chloral, for example, and benzedrine, the bromides, and the barbiturates.
>
> Most of these modifiers of consciousness cannot now be taken except under doctor's orders, or else illegally and at considerable risk. For unrestricted use the West has permitted only alcohol and tobacco. All the other chemical "doors in the wall" are labelled "dope," and their unauthorized takers are "fiends."

PP. 61–62

The decision by the United Nations Commission on Narcotic Drugs to declare benzodiazepines "controlled" substances was made in February 1984 after many years of deliberation and a number of dissenting votes.

CHAPTER 8

P. 63

As *Science* graphically made clear with its missing E, there is a tremendous literature on opiate receptors and endogenous opioids. To trace the history of this exciting development, three review articles by Solomon H. Snyder are recommended. The first appeared in the March 1, 1979, issue of *The New England Journal of Medicine*. The second appeared in the August 29, 1980, issue of *Science*. The latest was in the April 6, 1984, issue of *Science*. The discovery of endorphins is reported in the September 17, 1976, issue of *Science* in a paper by Avram Goldstein. The 1984 review by Snyder provides many of the references that space limitations prevent listing here.

P. 66

The finding that the brain contains substances that resemble morphine inspired the great American physician-essayist Lewis Thomas to speculate about the reasons why. He says he understands why humans might need something like morphine in their brains but why are the substances also found in brains of lower animals, including earthworms? "For a species as intelligent and at the same time as interdependent and watchful of each other as ours, it might be useful to install a device of this kind to guard against intolerable pain, or to ease the individual through what might otherwise be an agonizing process of dying. Without it, living in our kind of intimacy, at our close quarters, might be too difficult for us, and we might separate from one another, each trying life on his own, and the species would then, of course, collapse."

But why, he asks, should mice have the same equipment, and why, of all creatures, earthworms? Then he thinks about it and decides that maybe earthworms need a little morphine in their brains, also. "Without protection against overwhelming pain, the day-to-day life of a worm, being stepped on, snatched by birds, ground under plows, washed away in streams, would be hellish indeed."

Anyway, he says, "There it is [endorphins], a biologically universal act of mercy. I cannot explain it, except to say that I would have put it in had I been around at the very beginning, sitting as a member of a planning committee, say, and charged with the responsibility for organizing for the future a closed ecosystem, crowded with an infinite variety of life on this planet. No such system could possibly operate without pain, and pain receptors would have to be planned in detail for all sentient forms of life, plainly for their own protection and the avoidance of danger. But not limitless pain; this would have the effect of turbulence, unhinging the whole system in an agony even before it got under way. . . . I would have cast a vote for a modulator of pain . . . set with a governor of some kind, to make sure it never got out of hand. *The Youngest Science: Notes of a Medicine-watcher* by Lewis Thomas (Viking, 1983).

P. 71

The discovery of the benzodiazepine receptors is described in an article by P. Skolnick and S. M. Paul called "New concepts in the neurobiology of anxiety" in *The Journal of Clinical Psychiatry* 44 (II, sect. 2):12–19.

NOTES AND REFERENCES   205

P. 71

A recent article on the beta-carbolines appeared in the August, 1984, issue of *The Archives of General Psychiatry* in an article by T. R. Insel et al.

P. 72

Chloride plays an important role in the GABA-benzodiazepine relationship. A channel for chloride anions penetrates cell membranes. The diameter of the channel is regulated by nearby receptors for both GABA and benzodiazepines. It has been suggested that benzodiazepines relieve anxiety by collaborating with GABA to open the chloride channel faster and hold it open longer, thereby causing hyperpolarization of the cell.

P. 72

The miracles of PET scanning are described (among many other places) in the August 13, 1982, issue of *The Journal of The American Medical Association* and in an editorial in the January 3, 1985, issue of *The New England Journal of Medicine*.

Using the PET scanner, scientists at Washington University identified a discrete brain abnormality in patients with panic disorder (see Chapter 12). If reproduced by other investigators, this study will be of historical importance, demonstrating for the first time a definite correlation of abnormal brain activity with a psychiatric disorder. The study was published by E. M. Reiman, et al., and appeared in *Nature* 310:683–685.

P. 74

The classical paper by Pitts and McClure showing that anxiety attacks could be produced in patients with panic disorder by infusing them with sodium lactate appeared in *The New England Journal of Medicine* in 1967 (277:1329–1336). A recent article on the relationship of calcium to lactate infusion is in *Biological Psychiatry* (19:110–115).

P. 75

The locus ceruleus theory of anxiety is well described in *Panic Disorder and Agoraphobia* by Gorman, Liebowitz, and Klein, a 1984 publication by the Upjohn Company in Kalamazoo, Michigan.

CHAPTER 9

P. 80

Auenbrugger's description of nostalgia can be found in C. N. B. Camac's *Classics of Medicine and Surgery* (Dover Publications, 1959).

P. 80

Perhaps the best book ever written on the concept of disease was by Robert P. Hudson: *Disease and Its Control* (Greenwood Press, 1983).

P. 80

Tuke's Dictionary was published by J. and A. Churchill in 1892.

P. 81

The definition of disease was from D. W. Goodwin and S. B. Guze's *Psychiatric Diagnosis*, third edition (Oxford University Press, 1984).

P. 82

The freckles analogy was contributed by Robert P. Hudson (see above).

P. 83

Others echoed Freud's concern about the evangelistic fervor with which Americans embraced psychoanalysis. Karl Menninger wrote in 1955: "It must be recalled that Freud had always serious fears about the excessive popularity of psychoanalysis in America. He was afraid to see his essential goal diluted and compromised. Many of us who practice psychiatry and psychoanalytic teaching have partaken of this fear" (*Psychiatrie et Psychanalyse in La Psychiatrie Dans LeMonde Encyclopedie Medical-Chirurgicale*, Paris, 1955).

P. 83

Freud expressed his doubts about psychoanalysis as a therapy to J. Wortis (*Fragments of an Analysis With Freud*, Simon & Schuster, 1954). Freud's comment about Lourdes was made in his 1933 book, *New Introductory Lectures on Psychoanalysis* (Norton).

P. 83

Freud may have become disillusioned about the effectiveness of psychoanalysis late in life but for many years, like most analysts, he maintained that every successful analysis was evidence of the truth of the analyst's interpretation and indirect evidence of the validity of psychoanalytic theory. A recent book by Adolph Grünbaum (*The Foundations of Psychoanalysis: A Philosophical Critique*, University of California Press, 1985) maintains that therapeutic success cannot support the validity of analytic theories for two reasons: First, there was little systematic evidence that such success occurred. Second, when it did occur, there was no proof the patient might not have recovered without the analysis or with another treatment.

Grünbaum claims that the only way to test the validity of analytic theory is *outside* the clinic. The Freudian notion, for example, that paranoia is a manifestation of latent homosexuality can be tested "outside the clinic" by determining whether there has been a decline in paranoia concomitant with increased permissiveness toward homosexuality. (There is no evidence whether paranoia is increasing or decreasing, but older clinicians mostly agree that a decrease, if any, has been slight.)

P. 84

In the mid-1980s several books appeared purporting to present evidence that Freud's theory of the unconscious mind had been confirmed in laboratory tests. What actually had been shown was that the brain registers many impressions outside of conscious awareness. This has been known for a long time; hypnotic suggestion is perhaps all the proof one needs.

However, Freud's theory of an unconscious mind was not based on this simple observation. Rather, the unconscious mind was a storehouse of memories and desires actively kept out of conscious awareness through the "mechanism" of repression. According to the repression theory, which Freud called the foundation of psychoanalysis, the repressed memories were expressed symbolically in dreams and neurotic symptoms, but otherwise could not be recovered except by special techniques such as free association. This theory has *not* been subjected to scientific testing and indeed is probably untestable.

Other books published in 1985 attempted to reconcile Freudian theory with new knowledge about the brain. These included *Brain and Psyche* by Jonathan Winson (Anchor Press/Doubleday); *Mind, Brain, Body* by Morton F. Reiser (Basic Books); and *Starwave* by Fred Alan Wolf (Macmillan). Winson finds parallels between the unconscious mind and "off-line" computer processing. Reiser thinks the "pendulum has swung too far" toward biological psychiatry but still hopes for a convergence of biology and Freudian ideas in the future. At one point he compares a patient cured of a subway phobia through psychoanalysis with patients cured of phobias by drugs and implies that one can infer a similarity in the "basic mechanisms" when "similar results can be brought about by these very different treatment modalities." As Richard M. Restak, a reviewer of the book, pointed out, this is nonsense: "One deduces mechanisms from similar results at one's peril. A patient may exhibit paralysis because of a brain tumor, a stroke or hysteria. In each instance, the mechanisms are different." Wolf, a physicist, bases *his* convergence on quantum mechanics—an esoteric approach indeed!

P. 84

Freud summed up his life's work in *An Autobiographical Study* (Hogarth, 1935).

P. 84

In referring to his work as scientific, Freud rarely mentioned his belief in telepathy, Lamarkian inheritance, and—his favorite theory of all—the "death instinct."

P. 84

Not everyone agrees with Karl Popper's definition of science. As Jonathan Lieberson pointed out (*New York Review of Books*, January 31, 1985), "What is important for the scientific status of a theory is the promise it holds to contribute to a systematic and controlled knowledge of the world. Even a vague, untestable, or false theory can have such promise if it suggests an interesting path of research to scientists. Moreover, whether or not a theory is itself 'scientific,' it can be investigated in a scientific spirit. Even if it had no testable claims of its own, psychoanalysis could still be scientific in so far as psychoanalysts made efforts to clarify their hypotheses and submit them to rational criticism."

He concludes later that psychoanalysts have failed to do either.

PP. 84–85

Popper devotes pages 34–39 in his book *Conjectures and Refutations* to psychoanalysis, the source of his comments in this chapter (published by Basic Books in 1962).

P. 85

Freud's rather startling statement that psychoanalysis was a "postdictive science" was made in his paper on the "Psychogenesis of a case of homosexuality in a woman" (Collected Papers, II. London, 1933).

PP. 85–86

Klein's litany of largely discarded theories by prominent psychoanalysts was included in a formal debate sponsored by The American Psychiatric Association at its annual meeting in Los Angeles in 1984. Besides Klein, the participants were myself, John Nemiah, Robert Michels, and Byram Karasu (moderator).

P. 86

Freud himself expressed doubt about whether paranoia was *always* due to latent homosexuality. He had examined a paranoid patient and was unable to find elements of latent homosexuality. Freud should have been the first to realize, however, that this in no way disproved the theory. The patient's repression of homosexual impulses may have been too strong for even Freud to overcome. This is one more example of a Freudian theory that cannot be falsified.

P. 87

Ernest Jones' "unexpected conclusion that hardly any of Freud's early ideas were completely new" appeared in an article in *The Journal of Medical Science* (100:198–210).

P. 87

Bailey's scathing denouncement of psychoanalysis was later published in the November 1956 issue of *The American Journal of Psychiatry*.

P. 88

Jones' declaration that he had found a religion in psychoanalysis is included in an article in *The International Journal of Psychoanalysis* (27:76–81).

P. 90

Even the New York literati may be losing the faith. Recent issues of *The New York Review of Books* and *The New Yorker* have published articles by Janet Malcolm that have been utterly devastating to the cause of psychoanalysis (e.g., the December 5 and 12, 1983, issues of *The New Yorker*; the December 20, 1984, issue of the *NYROB*).

P. 91

The passing of the guard from psychoanalysis to "biological" psychiatry was clearly evident in the 1984 membership of the psychiatry section of the National Board of Medical Examiners. Most medical students in the United States must pass "National Boards" to receive an M.D. degree or a license to practice. Between 1960 and 1984 there was a conspicuous change in the philosophical orientation of the members of the psychiatry section. In 1984, seven of the eight members were clearly of the biological school; 25 years previously psychiatrists of psychoanalytic persuasion dominated the section. And *these* are the people who select the questions that determine whether medical students are permitted to practice medicine!

P. 91

The last nail in the analytic coffin (the metaphor admittedly is premature) may have been the recent finding by a malpractice tribunal that a private psychiatric facility specializing in psychoanalysis had been negligent in withholding drugs from a patient and providing only psychotherapy. As Alan A. Stone, Professor at Harvard Law School, a psychiatrist and former President of The American Psychiatric Association, wrote in the November 22, 1984, issue of *The New England Journal of Medicine*: "Treated by the psychodynamic approach in a specialized private hospital, with an excellent psychodynamically trained staff, the patient did not recover. Within weeks of his transfer to another hospital and the initiation of biological treatment with antidepressants, he had a dramatic recovery. His malpractice suit speaks to the efficacy of biological psychiatry and the inefficacy of psychodynamic psychiatry, even when the latter is provided by highly trained and skilled clinicians.

"Indeed, the patient's expert witnesses, including highly distinguished members of the academic community, argued that the private facility had been negligent not only in its failure to administer biological treatment but also in not providing staff psychiatrists with biological training. They questioned the basic validity of a psychiatric hospital specializing in psychodynamic approaches" ["psychodynamic" is more or less synonymous with psychoanalytic].

Stone asked: "Has psychiatry reached the point where use of the psychodynamic model is viewed as malpractice when it is the exclusive treatment for serious mental disorders?" He didn't answer, but the patient was awarded $250,000 by the tribunal. (At this writing the case had not gone to trial.)

PP. 91–92

The P. D. Scott quote is from a book review in *The British Journal of Psychiatry* (113:929–930).

P. 92

Members of the APA Task Force, besides Spitzer, were Nancy Andreasen, Robert L. Arnstein, Dennis Cantwell*, Paula J. Clayton*, William A. Frosch, Donald W. Goodwin*, Donald F. Klein, Z. J. Lipowski, Michael L. Mavroidis, Henry Pinsker, George Saslow*, Michael Sheehy, Robert Woodruff (deceased)* and Lyman C. Wynne. Asterisks refer to members who had associations with Washington University.

P. 93

DSM-III was not only a best seller in America but had an unexpectedly profound effect on psychiatrists throughout the world. A book called *International Perspective on DSM-III* (edited by Spitzer, Williams, and Skodol; American Psychiatric Association, 1983) contained critiques of DSM-III by well-known psychiatrists from around the world, including not only countries in Western Europe but also from Asia and the Third World. They were generally favorable. Interestingly, the Chinese Peoples Republic liked the book best, while the most lukewarm response came from the French.

P. 94

For a lucid, engaging, and humane treatment of the biological revolution in psychiatry, no book can be more highly recommended than Nancy Andreasen's *The Broken Brain*, published in 1984 by Harper & Row.

NOTES AND REFERENCES 211

CHAPTER 10

P. 97

DSM-III refers to the *Diagnostic and Statistical Manual of Mental Disorders* (third edition), published by The American Psychiatric Association in 1980. The introduction by Robert L. Spitzer, Chairman of the DSM-III task force, describes in more detail the "rules" for diagnosis presented in this chapter.

PP. 97–99

A critical examination of current psychiatric classifications of anxiety was recently published in *The British Journal of Psychiatry* (144:78–83). As the author, Peter Tyrer, points out, the American classification differs from the European system in that panic is given separate diagnostic status in America whereas it is merely a symptom in European systems.

CHAPTER 11

P. 101

Generalized anxiety disorder is a term introduced by DSM-III. In the past it has had a variety of names, including neurasthenia, hypochondriacal affections, depressions of spirit, the blues, or chronic anxiety reaction. In the eighteenth and nineteenth centuries a favorite term was "vapors," which both describes generalized anxiety and fainting fits (probably brought on by hyperventilation). In Victorian times the prototype of a refined young woman was a "swooner, pale and trembling, who responded to unpleasant or unusual social situations by taking to the floor in a graceful and delicious maneuver, in no way resembling the crash of the epileptic." A Jane Austen heroine found one social situation "too pathetic for the feelings of Sophie and myself. We fainted alternately on a sofa." Overtight corsets may have been responsible for some of the fainting. A nineteenth-century physician, Dr. John Brown, cured fainting by "cutting the stay laces, which ran before the knife and cracked like a bow string" (cited by Goodwin and Guze in *Psychiatric Diagnosis*, 3rd edition, Oxford, 1984).

PP. 102–105

E. B. White's symptoms are described in his biography by Scott Elledge (Norton, 1984).

P. 105

Generalized anxiety disorder was compared with panic disorder in a recent paper by Anderson, Noyes, and Crowe (*American Journal of Psychiatry*, April, 1984). Subjects with generalized anxiety disorder were shown to have fewer autonomic symptoms and an earlier, more gradual onset. The illness was also observed to have a more chronic course and higher rate of improvement over a period of many years.

P. 105

In discussing anxiety one must distinguish between *trait anxiety* and *state anxiety*. Trait anxiety is "anxiety proneness." We all differ in our predisposition to perceive our environment as dangerous or threatening, and these differences are presumably lifelong. By contrast, state anxiety refers to one's feelings here and now, at a precise moment. A person low in trait anxiety may occasionally experience high state anxiety, but someone who is highly trait-anxious will experience more and intenser movements of anxiety. Individuals suffering from depression are often high in trait anxiety. The 27-year-old electrician with symptoms of only two years' duration may indeed have been diagnosed as having a clinical depression by some psychiatrists. (From the article on anxiety by Charles Spielberger in *The ABC of Psychology*, L. Kristal, ed. Pelican Books, 1981).

PP. 107–108

In the lead editorial in the January 14, 1984, issue of *The Lancet*, the question of whether psychotherapy is an "effective treatment or expensive placebo" is examined. The journal points out that although the concept of talking to patients to help them overcome psychological distress is as old as language itself, psychotherapy as a medical discipline is much more recent. It began with Mesmer and achieved respectability and "sometimes worship" during the lifetime of Freud.

"In the heady days of psychotherapy early this century, few questioned the efficacy of psychotherapy and little thought was directed toward evaluating its outcome. Even when other effective treatments such as drug and behavior therapy appeared on the scene and were shown to be effective, the adherents of psychotherapy were reluctant to evaluate their treatment in the same way. Many judged this unnecessary; psychotherapy was so much part of psychiatry that it did not need evaluation. . . . This state of affairs could not last, and when science finally caught up with psychiatry, psychotherapy could no longer escape proper evaluation." Hans Eysenck was the first to challenge the effectiveness of psychotherapy by comparing the improvement rates of patients who were seen in psychotherapy and other groups of subjects who were not patients but had similar symptoms ("The effects of psychotherapy: an evaluation." *Journal of Consulting Psychology* 16:319–324). The two groups did about equally well and Eysenck concluded that psychotherapy had nothing to offer over the rate of spontaneous remission. Argument has raged ever since but published reports are hard to compare because of great differences in methodology.

In 1980, an analysis of nearly 500 studies comparing psychological treatment with a control group suggested that psychotherapy was significantly superior (Smith, Glass, Miller: *The benefit of psychotherapy*. Johns Hopkins University

NOTES AND REFERENCES   213

Press, 1980). However, in this analysis placebo therapy was regarded as a form of psychotherapy and not included with the controls. The patients receiving placebo therapy differ from untreated controls in that they believe they are receiving an active treatment and therefore expect to get better. Other investigators reanalyzed the data from the studies in which psychotherapy was compared with placebo treatment and found no difference between the outcomes of psychotherapy and placebo treatment. The data indicated that some psychotherapists help patients to improve while others make them worse.

There is general agreement that the selection of patients for psychotherapy is crucially important. As J. Rabkin writes, "Patients who are young, physically attractive, well-educated members of the upper middle-class, intelligent, verbal, willing to talk about and have some responsibility for their problems, and show no signs of gross pathology are welcomed as good therapeutic candidates." (J. Rabkin: "Therapeutic attitudes toward mental health and illness," A. S. Gurman and A. M. Razin, eds. *Effective Psychotherapy: A Handbook of Research*, Pergamon, 1977). According to H. Stupp, such patients show the greatest levels of spontaneous improvement and the efforts of the psychotherapist add little (H. Stupp: "The outcome problem in psychotherapy revisited" (*Psychotherapeutic Theory, Research and Practice*, 1:1–13).

The Lancet ended the editorial by saying that formal evaluation of psychotherapy is essential. "If nothing is done psychotherapy will doubtless survive but will drift even more into the fringes of medicine, an option only for those who are prepared to pay and ask no questions about outcome."

CHAPTER 12

PP. 112–123

The Paul Dudley White study was published in the *Journal of the American Medical Association* in 1950 (142:878–889).

DeCosta is credited with the first description of panic disorder, which appeared in the *American Journal of Medical Science* in 1871 (61:17). He called the condition "irritable heart," but it quickly became known as DeCosta's syndrome.

P. 113

Panic disorder patients suffer from a multitude of symptoms other than apprehension. The frequency of occurrence of these symptoms can be found in a paper by M. Cohen and P. White cited in *Psychiatric Diagnosis*, third edition (Oxford, 1984) by D. W. Goodwin and S. B. Guze.

P. 115

The early onset of panic disorder is well documented in numerous papers, including those by G. Winokur and E. Holeman (*Acta Psychiatrica Scandinavia*

39:384–412) and R. A. Woodruff, Jr., S. B. Guze and P. J. Clayton (*Comprehensive Psychiatry* 13:165–170).

PP. 115–116

The relationship of irritable colon to panic disorder is described in a paper by J. Liss, D. Alpers, and R. A. Woodruff in *Diseases of the Nervous System* (34:151–157).

P. 116

Paradoxically, *raising* blood levels of carbon dioxide actually produces anxiety symptoms in patients with panic disorder. Symptoms can be prevented by pretreating the patients with antidepressant medication (see Chapter 18). (M. A. Vandenhout and E. Griez, "Panic symptoms after inhalation of carbon dioxide," *British Journal of Psychiatry* 144:503–507). As noted in Chapter 8, intravenous infusions of sodium lactate will also produce panic attacks in victims of the disorder. These also are prevented by premedication with antidepressants ("Lactate provocation of panic attacks" by M. R. Liebowitz, et al., *Archives of General Psychiatry*, 41:764–770).

P. 116

David Sheehan's book, *The Anxiety Disease*, was published by Charles Scribner's Sons in 1983. It contains many vivid descriptions of anxiety symptoms.

P. 117

The literature on mitral valve prolapse has become voluminous in recent years. An excellent review of the evolving concept of mitral valve prolapse can be found in the August 5, 1982, issue of *The New England Journal of Medicine*.

P. 119

Depression is the most common complication of panic disorder. Of patients examined in medical clinics of a general hospital, 34% report episodes of depression (J. Clancy, et al., *Journal of Nervous and Mental Disease* 166:846–850). Because of the frequent occurrence of depression in panic disorder patients, there has been speculation that panic disorder is a variant of primary depression. Both conditions are relieved by antidepressant medication. Recent evidence indicates that there is an overlap of depression and panic disorder in the relatives of depressed patients (Leckman, et al., *American Journal of Psychiatry* 140:880–882). Arguments against the illnesses being identical are as follows: (a) primary depression is associated with a high rate of suicide and panic disorder is not; (b) panic disorder has an earlier age of onset than primary depression; (c) panic attacks can usually be precipitated in people with a history of spontaneous attacks by the intravenous infusion of sodium lactate, and this rarely happens in depressed patients.

NOTES AND REFERENCES 215

P. 120

A review of anxiety disorders and alcohol abuse is published in the September 1984 issue of *The Journal of Studies on Alcohol* and also in the January 1985 issue of *The Journal of Clinical Psychiatry*.

P. 120

Some argue that panic disorder and generalized anxiety disorder are the same illnesses. However, when the family members of patients with panic disorder are studied, there appears to be no increase of generalized anxiety disorder (or any other anxiety disorder) in the families. This would suggest that panic disorder is a familial disease not associated with an increased familial risk of other psychiatric conditions (R. Crowe, et al., *Archives of General Psychiatry* 40:1065–1069). Twin studies support the likelihood that panic disorder is a separate condition from other anxiety disorders. Panic attacks were five times more frequent in identical-twin pairs than in fraternal-twin pairs in one study, whereas generalized anxiety disorder was equally distributed between identical and fraternal twins (S. Torgersen: *Archives of General Psychiatry* 40:1085–1089).

P. 122

Regarding the assertion that supportive psychotherapy is of "great importance" in treating panic disorder, a 1984 report by H. Garakani, C. Citrin, and D. F. Klein (*American Journal of Psychiatry* 141:446–448) reported that ten patients with panic disorder were treated solely with imipramine (an antidepressant) without psychotherapy. All reported cessation of panic attacks. Most therapists, however, still combine medication with psychotherapy.

CHAPTER 13

P. 124

The phobic psychiatrist is found in the chapter on phobic disorder by John C. Nemiah in *Comprehensive Textbook of Psychiatry*, third edition, edited by Kaplan, Freedman, and Sadock (Williams & Wilkins, 1980).

P. 125

The survey by the National Institute of Mental Health was published in the October 1984 issue of the *Archives of General Psychiatry*.

P. 125

The study comparing patients with predominant mood disorders and patients from the fracture clinic was conducted by Shapiro, Kerr, and Roth (*British Journal of Psychiatry* 117, 25–32).

P. 125

The unexpectedly high rate of agoraphobia and social phobia in alcoholics was reported by Mullaney and Trippett in the *British Journal of Psychiatry* 135, 565–573. Terhune also reported that phobics were susceptible to alcohol and drug dependence (*Archives of Neurology and Psychiatry* 62, 162–172), but Sim and Houghton found a low rate of alcohol and drug dependence in phobics (*Journal of Nervous and Mental Diseases* 143, 484–491), leaving the issue unresolved.

P. 126

Fears and Phobias by I. M. Marks (Academic Press, 1969) is the source for the brief historical remarks.

P. 126

The paper by Karl Westphal was called "Die Agoraphobie: eine neuropathische Erscheinung, *Arch. fur Psychiatrie und Nervenkrankheiten* 3:138–171, 219–221, 1871–1872.

P. 127

Of the 13 million American adults suffering from anxiety disorders, nearly 11 million have phobic disorders. This accounts for the inclusion in this book of four separate chapters on four separate phobic disorders, plus a chapter on the treatment of phobia. A source much relied on was Isaac M. Marks, M.D. of the Institute of Psychiatry at the Maudsley Hospital in London. Dr. Marks is perhaps the world's leading authority on phobias (his closest competitor, in the United States, is Donald F. Klein). Marks' book, *Fears and Phobias*, published in 1969, provided invaluable background material for the chapters on phobic disorders. Dr. Marks subsequently has published widely in the fields of fear and phobia. When Marks is mentioned in the text or in these notes, the source of information was *Fears and Phobias*. References to his other writings are given in full each time they appear.

CHAPTER 14

P. 128

Airplane phobics have devised ingenious ways to conquer their phobia. One man, for example, learned that the probability of *two* bombs being hidden in an airplane was one in a billion. To cure his phobia he always carried a bomb with him on airplanes (story probably apocryphal).

Recommended reading for this chapter is *Phobia: Psychological and Pharmacological Treatment*, edited by Mavissakalian and Barlow (Guilford Press, 1981)

NOTES AND REFERENCES

and, of course, the classical work of Marks described in the Notes for Chapter 13.

Other books used for background in this chapter included *Phobias and Obsessions* by Melville (McCann & Geoghegan, 1977) and *Phobias: Their Nature and Control* by Rachman (Thomas, 1968).

P. 129

The groundbreaking book by Joseph Wolpe was called *Psychotherapy by Reciprocal Inhibition* (Stanford University Press, 1958).

P. 130

One reason for so many fancy names for phobias is that doctors have long recognized that fancy names are comforting and even therapeutic. The psychiatrist E. Fuller Torrey compared witch doctors with psychiatrists. Both, he said, make you feel better because of the certainty exuded by the "authority figure, the diploma on the wall, or the proper headdress, bones and rattles, and finally, because the authority in each case gave the condition a name" (from *The Mind Game*, Bantam, 1973). You have a *curse* from your dead mother-in-law, or you have a *bug* that's going around. Torrey called it the Rumpelstiltskin effect. Rumpelstiltskin was a dwarf who helped a young woman become queen if she would give him her first-born child. When she asked for the child back, he refused because she didn't know his name. If she learned his name in three days, he said, she could keep the child. At midnight on the third day there is a scene where the queen says archly, "Is your name Michael? John? Rumpelstiltskin?" And poor Rumpelstiltskin goes pop! Vaporizes. Disappears. If you can give it a name, it will disappear (Adam Smith's version in *Powers of Mind*, Random House, 1975).

P. 133

Freud's discussion of animal phobia appeared in *Totem and Taboo* (Hogarth Press, 1955).

P. 133

Why do some people experience fear looking *up* at a high building as well as looking down from one? There has never been a good answer.

P. 134

Variations on the theme of heights and the notion that a fear of heights is really a fear of falling come from Isaac Marks.

P. 134

MacAlpine has an excellent article on syphilophobia in the *British Journal of Venereal Diseases* 33, 92–99.

P. 135

Rogerson describes how a man stopped worrying about having syphilis once he found he actually had it (*British Journal of Veneral Diseases* 27, 158–159).

P. 135

Ladee has an excellent review on hypochondriasis called *Hypochondriacal Syndromes* (Elsevier, 1966).

P. 136

Isaac Marks presented the data on the prevalence of illness phobia.

PP. 136–137

The three follow-up studies of phobia are "Phobic disorders four years after treatment: a prospective follow-up," by Marks (*British Journal of Psychiatry* 118, 683–688); "The natural history of phobia" by Agras, Chapin, and Oliveau (*Archives of General Psychiatry* 26, 315–317); "Treatment of phobias" by Klein, Citron, Woerner, and Ross (*Archives of General Psychiatry* 40, 139–45).

CHAPTER 15

P. 138

The reader can try a little experiment to demonstrate the power of staring. Stand on a street corner where cars pull up at a street sign. Stare at some drivers and not at others. Those you stare at will move away from the street sign faster than those you ignore. (Sometimes, anyway.)

Those super-polite people, the Japanese, seem to have more than their share of social phobias. These are lumped under the name "shinkeishitsu" and include a fear of blushing and eye-to-eye contact; a fear of emitting bad odors; and a fear that one's facial expression may annoy others (Caudill and Schooler, *Mental Health Research in Asia and Pacific*, East-West Center Press, 1969).

PP. 139–140

One of our first public performances involves the act of elimination. Successful performance is rewarded with smiles and congratulations. The soiled diaper, on the other hand, elicits a reaction the child only learns the name for later: disgust.

Presumably children are not born with disgust toward their bodily products, but the force of the mother's response assures the early acquisition of this interesting response to such things. By age three or four, most children are as disgusted by feces as their mothers are, and have discovered bad smells from other sources, such as vomit. (Negative reactions to certain odors may have survival value and therefore a genetic basis; spoiled food is as noxious to many animals as to people. Our culture, however, has extended the range and subtlety of offensive odors to an unprecedented degree, creating a huge market for deodorants.)

Freudians make much of toilet training as a source of later adult ambivalence: whether to go or not to go. Indecisive people are supposed to have had unfortunate toilet training experiences. Some Freudians also believe the act of "going" is early identified by the child as aggressive and therefore potentially useful in combating a hostile environment. The association of excretory processes and aggression seems borne out by the choice of some well-known four-letter words as terms of abuse. According to the anthropologist Margaret Mead, every society has a stock of words based on excrement and sexual acts to express anger.

Another Freudian notion is that the act of urination, at least by men, becomes confused at an early age with competition with one's peers. It is true that boys will sometimes compete to see who can urinate the furthest. One twist of this quaint sport (if mothers only knew!) consists of writing words in the snow with urine tracks, with some boys showing more skill than others.

For most young boys this is a "phase" they pass through quickly, but a few become obsessed with competition on this primitive level, resulting in what Freudians call a "Urethral Personality" (later renamed Type A personality).

P. 140

A recent American President was said to dictate to his secretaries (both sexes) while seated on a commode. The President may have been counterphobic (see p. 44).

P. 140

One reason women may have more urinary tract infections than men is that they develop stretched bladders from delaying urination. Many little girls will go to school and not urinate until they return home, sometimes because of a phobia about sitting on strange toilet seats. Another explanation unrelated to phobias is that women take more baths than men and have a shorter distance for germs to travel to reach the bladder.

P. 141

The standard textbook for treating social-sexual phobias is *The New Sex Therapy* by Helen Singer Kaplan (Brunner/Mazel, 1974).

P. 142

Hippocrates' description of crowd phobia is found in Robert Burton's *The Anatomy of Melancholy*.

CHAPTER 16

P. 145

A book by Mathews, Gelder, and Johnston called *Agoraphobia* (Guilford Press, 1981) is the most comprehensive work on the subject. It is heavily referenced and contains a complete description of a self-help treatment method. Gelder worked with Isaac Marks, whose book, *Fears and Phobias*, was also used as a resource for this chapter.

PP. 145–146

Along with *Platzschwindel* and barber's chair syndrome, agoraphobia has a host of synonyms, including phobic-anxiety-depersonalization syndrome, locomotor anxiety, street fear, phobic anxiety state, and nonspecific insecurity fears.

P. 146

Donald Klein and other authorities believe that fear of crowded places and open spaces is really secondary to a dread of being away from home where help cannot be readily obtained in the event a panic attack occurs. Like Freud, they believe that agoraphobia is almost always ushered in by a panic attack that is so frightening that the person wishes fervently to avoid a recurrence, and this leads to the many phobic avoidances that characterize the disorder.

PP. 147–148

Vivid specimens of agroaphobic behavior can be found in Marks' 1970 paper, "Agoraphobic syndrome" in the *Archives of General Psychiatry* 23, 538–553; *Agoraphobia in the Light of Ego Psychology* by Weiss (Grune & Stratton, 1964); "A long-term follow-up study of neurotic phobic patients in a psychiatric clinic" by Errera and Coleman in *The Journal of Nervous and Mental Diseases* 136, 267–271; and "Confinement in the production of human neuroses: the barber's chair syndrome" in *Behavioral Research and Therapy* 1, 175–183.

PP. 149–150

See Notes for Chapter 15 for studies on phobia and alcoholism. Further evidence that agoraphobia is not a variant form of depression was found in a recent study in Iowa. Of the immediate relatives of agoraphobics, 9% had agoraphobia, but there was less depression in the relatives of agoraphobics than in nonpsychiatric patients. The study provided further evidence of a connection between agoraphobia and alcoholism: 35% of the fathers of agoraphobics were alcoholic compared to 10% of fathers of nonpsychiatric patients. ("A family study of agoraphobia" by Harris, Noyes, Crowe, and Shaudhry, *Archives of General Psychiatry* 40:1061–1064.

CHAPTER 17

P. 152

A good paper on school phobia is by L. Hersov (*Journal of Child Psychology* 1, 130–136). Many of the comments about the problem came from this source as well as from Isaac Marks' book, *Fears and Phobias*.

P. 154

Billy's school phobia is described in an article by R. E. Smith and T. M. Sharpe in *The Journal of Consulting and Clinical Psychology* 35, 239–243.

P. 155

Reluctant mothers are well described in Eisenberg's paper on school phobia in *The American Journal of Psychiatry* 114, 712–718.

P. 156

Seventy-one school-phobic children seen at a child guidance clinic were followed up after they passed school-leaving age. Most were doing well ("School phobic children at work," by H. Baker and U. Wills, *British Journal of Psychiatry* 135, 561–564.

P. 156

Donald Klein and his wife, Rachel Gittelman-Klein, found that imipramine, an antidepressant drug, was useful in relieving panic attacks in patients with agoraphobia. They proceeded to see whether the drug also relieved symptoms of school phobia, and it did. ("School phobia: diagnostic considerations in the light of imipramine effects," *Journal of Nervous and Mental Diseases* 156, 199–215.)

CHAPTER 18

P. 157

Two excellent recent books on the treatment of phobia are *Phobia* edited by Mavissakalian and Barlow (Guilford Press, 1981) and *Agoraphobia* by Mathews, Gelder, and Johnston (Guilford Press, 1981). Also recommended are *Phobia: A Comprehensive Summary of Modern Treatments* by DuPont (Brunner/Mazel, 1982); *The Behavioral Management of Anxiety, Depression and Pain* edited by Davidson (Brunner/Mazel, 1976); and *Anxiety: Its Components, Development, and Treatment* by Lesse (Grune & Stratton, 1970).

Books on fear and phobia written for the lay reader include *Living with Fear: Understanding and Coping with Anxiety* by Marks (McGraw-Hill, 1978); *Phobias and Coming to Terms with Them* by Emerson (British Medical Association, 1969); and *How To Master Your Fears* by Steincrohn (Funk, 1952).

Wolpe's book was published by Stanford Press in 1958. Evidence that reciprocal inhibition is effective for simple phobias is presented in a paper by McCononaghy in the *British Journal of Psychiatry* 117, 89–92 and reviewed in the book edited by Mavissakalian and Barlow (see above).

P. 159

Jacobson's book on progressive relaxation was published by the University of Chicago Press in 1938.

A more recent book by Jacobson is *Anxiety and Tension Control* (Lippincott, 1964).

P. 165

The strongest evidence that long-term use of benzodiazepine tranquilizers results in withdrawal effects is presented in a paper by Malcolm Lader in the *Journal of Clinical Psychiatry*, April, 1983. The most common withdrawal symptoms were anxiety, tension, and sleep disturbance, but a small minority of patients also had more serious withdrawal effects, including paranoid reactions and visual hallucinations. The withdrawal symptoms usually went away within two to four weeks. Lader also reported data suggesting that tolerance does not develop to therapeutic doses of benzodiazepines (i.e., the drugs do not lose their effects even after several weeks of taking them daily).

P. 165

Studies indicating that MAO inhibitors relieve anxiety states, including phobias, are reviewed by Grunhaus, Gloger, and Weisstub in *The Journal of Nervous and Mental Disease* 169, 608–613.

P. 168

Two articles appeared in the February 1983 issue of the *Archives of General Psychiatry* presenting the latest evidence that drugs relieve certain aspects of phobias. The two articles, however, provide somewhat different conclusions. A New York group headed by Donald Klein reported that imipramine prevented panic attacks in phobic patients, although it did not relieve anticipatory anxiety. (Klein elsewhere recommends benzodiazepines for anticipatory anxiety.) Klein also reported that supportive psychotherapy (consisting mainly of reassurance) was as effective as behavior therapy, as long as both led to confrontation with the phobic situation. In contrast, a British group headed by Isaac Marks concluded that imipramine had no effect on panic attacks, but "in vivo self-exposure" was a potent treatment for agoraphobia. The Klein group gave higher doses of imipramine and studied the results while the patients were still taking the drug, whereas the Marks group evaluated the drug effects after the drug had been terminated. As mentioned in this chapter, drugs only seem to be useful in relieving phobia while they are being taken.

PP. 168–169

The improvement in musical performance ascribed to propranolol was reported in the September 12, 1982 issue of the *New York Times*.

CHAPTER 19

PP. 172–176

Goodwin and Guze in *Psychiatric Diagnosis* (Oxford University Press, 1984) reviewed the findings of 13 follow-up studies of individuals with obsessive complusive disorder. The observations in this chapter on age of onset, course, and outcome of the illness are based on these follow-up studies, which show a remarkable degree of consistency in their findings. The studies were conducted in six different countries.

PP. 172–174

Common symptoms of obsessional illness are described by N. L. Gittleson in *The British Journal of Psychiatry* 112:261–264. The Karl Jasper's quote is from his monumental *General Psychopathology* (University of Chicago Press, 1963).

P. 174

Mrs. X was described in an Associated Press feature article in January 1984.

P. 174

Rituals are described in "Phenomenology of obsessive-complusive neuroses" by R. Stern and J. Cobb in *The British Journal of Psychiatry* 132:233–239.

P. 175

Prognostic features of obsessive complusive disorder are described in "Natural history of obsessional states: A study of 150 cases" by J. Pollitt in *The British Medical Journal* 1:194–198. The warning about "insight" therapy making the patient worse is in an article by L. Salzman and F. H. Thaler in *The American Journal of Psychiatry* 138:3, 286–296.

PP. 176–177

The studies indicating that behavior therapy produces relief from complusive rituals are reviewed by I. Marks, R. Hodgson and S. Rachman in an article in *The British Journal of Psychiatry* 127:349–364. Marks reviewed the subject again in a paper in *The American Journal of Psychiatry* 138:5, 584–592.

P. 177

The effectiveness of chlorimipramine was reported by Marks, et al. (among others) in two articles in *The British Journal of Psychiatry* (136:1–25, 161–166).

CHAPTER 20

P. 178

I am grateful to Dr. William E. Rinck, Director of the Psychiatric Outpatient Clinic at the Veterans Administration Medical Center in Kansas City, Missouri, for sharing with me his broad knowledge of the post-traumatic stress disorder.

P. 179

The London surgeon who challenged Erichsen's theory was employed by the London and Northwest Railroad and may not have been totally unbiased.

P. 179

An early report of nostalgia was contained in the United States Surgeon General's report of 1868 and apparently included malingering as well as PTSD.

PP. 179-180

The three principles for treating shell shock were proposed by a select committee advising the Surgeon General on the "formidable problem of war neurosis." Oddly, during World War II, the proposals had been overlooked and new committees were formed to study the problem. The importance of on-site treatment and social support was again recognized. However, these principles were not implemented until the Korean War. Success with rapid treatment near the front in Korea led to the current principles of combat psychiatry of Immediacy, Proximity, and Expectancy, first used on a large scale in Vietnam fifty years after the principals had been solidly documented in a select committees's report to the Attorney General.

PP. 178-183

I am grateful to Dr. Rinck for a tape recording of the Second National Conference on the Treatment of PTSD in Chicago in October 1983. The keynote address by Charles Figley, "Toward a generic view of traumatic stress," was particularly helpful in tracing the history of attitudes toward combat stress. Various punitive measures toward combating PTSD were described by J. R. Smith at the Second National Conference. The observations about the Vietnam War were also made at the conference.

For information about prisoners of war and concentration camp victims see J. Tas' paper in *The Psychiatric Quarterly* 25:679-685.

A basic work about PTSD is *Stress Disorders Among Vietnam Veterans* (Brunner/Mazel, 1978). The article by R. Grinker and J. Spiegel called "Men under stress" is particularly useful.

NOTES AND REFERENCES 225

PP. 182–183

The historical references were from *The Golden Bough* by H. G. Frazer, third edition, Macmillan, 1916 (Volume 3).

PP. 183–184

Symptom development in Vietnam veterans is described by V. J. DeFazio, S. Rustin and A. Diamond in *The American Journal of Orthopsychiatry* 45:158–170.

P. 185

In the August 1984 issue of *The American Journal of Psychiatry* veterans who did *not* develop post-traumatic stress after intense combat were studied and the authors found the following features: (1) calmness under pressure, (2) intellectual control, (3) acceptance of fear, (4) a lack of excessively violent or guilt-arousing behavior.

P. 185

Dr. Andreasen's comments are included in her article on post-traumatic stress disorder in *Comprehensive Textbook of Psychiatry*, fourth edition, H. I. Kaplan and B. J. Sadock, eds. (William & Wilkins, 1985).

P. 185

Predisposing factors to PTSD are described in a study in the January 1985 issue of *The American Journal of Psychiatry*. A history of familial psychiatric illness was found in 66% of the patients. Aside from PTSD, every patient had experienced at least one significant psychiatric illness during his lifetime, most commonly alcohol abuse or depression.

P. 185

The July 1983 issue of *Military Medicine* contains an article by R. H. Rahe and E. Gender that minutely traces the evolution of symptoms after acute wartime stress, particularly stress associated with captivity.

PP. 186–188

Compensation neurosis is described by M. H. Miller in *The Journal of Forensic Science* 4:159–170. Herbert Modlin, an authority on post-accident anxiety syndomes, also touches on compensation neurosis in an article in *The American Journal of Psychiatry* 123:1008–1012. Another useful article on the subject was written by T. L. Sonait and appeared in *The Southern Medical Journal* 53:365–375.

PP. 187–188

Katherine Quinn's advice for verifying a patient's symptoms was presented at the annual meeting of The American Psychiatric Association in Los Angeles in 1984.

P. 188

Robert Jay Lifton describes his experiences with Vietnam veterans in *Commonwealth* 91:554–556. His book *Home From the War*, published by Simon & Schuster in 1973, is an eloquent description of the plight of the Vietnam veteran.

P. 189

The literature by now on the post-traumatic stress disorder is so voluminous that interested persons are encouraged to write the National Library of Medicine for a literature search on the subject. Numbers 84-13, January 1982 through June 1984, were used in preparation of this chapter.

INDEX

A

"Abreaction," 44, 199n
Adler, Alfred, 84–85
Adrenaline, and emotion, 18–19
The Age of Anxiety (Auden), 12
Age of Faith, 11
Age of Reason, 11–12
Aged, benzodiazepine sensitivity, 56
Aggression, and benzodiazepines, 202n
Agonists
  benzodiazepines, 71
  definition, 66
  and drug development, 76–77
Agoraphobia, 145–51, 220n
  and alcoholism, 150, 216n, 220n
  breathing techniques, 169–70
  case histories, 220n
  versus crowd phobia, 142
  depression, 150, 220n
  diagnostic criteria, 146
  drug therapy, 164, 166–67, 222n
  history, 126
  prevalence, 125
Agoraphobia (Mathews, Gelder, and Johnston), 220n
Airplane phobics, 164, 216n
Alcohol
  antianxiety effect, 57
  and benzodiazepine mechanisms, 72
  conflict reduction, 36–37

Alcoholism
  and agoraphobia, 220n
  and anxiety disorders, review, 215n–16n
  and panic disorder, 120
All Faithful People (Lynd and Lynd), 194n
Alprazolam, 121, 164
America, and Freud, 86–87
Andreasen, Nancy, 210n
Animal phobias, 132–33
Antagonists
  benzodiazepines, 71
  definition, 66
  and drug development, 76–77
  opiates, 68–69
Antianxiety drugs. *See also* Benzodiazepines
  in generalized anxiety disorder, 106
  invention of, 51–62
  phobia treatment, 163–65
  safety, 60–62
Anticipatory anxiety, 129, 150
  drug treatment, 164, 166
Antidepressant drugs, 107, 121. *See also* Tricyclic antidepressants
  panic disorder, 121, 166
  phobic disorders, 165–68
The Anxiety Disease (Sheehan), 110, 116–17
Anxiety disorders. *See also specific disorders*

Anxiety disorders *(continued)*
definition question, 3–6, 191n–92n
DSM-III, 97–100
Apprehensive expectation, 102
Approach-avoidance studies, 30, 36
Autonomic hyperactivity, 20–21, 102
Autonomic nervous system, 18
Avoidance, and phobias, 123

B

Barbiturates
antianxiety effects, 57
and benzodiazepine mechanism, 72
conflict reduction, 36–37
Behavior therapy
versus drugs, agoraphobia, 166
in obsessive-compulsive disorder, 176–77, 223n
in phobias, 158–63
Behaviorism, 40–41
The Benzodiazepine Story (Sternbach), 201n
Benzodiazepines, 51–62
actions of, 58
aggression effect, 202n
in anticipatory anxiety, 222n
invention, 51–62, 201n–202n, 222n
in phobic anxiety, 163–65, 222n
receptors, 70–72, 204n
safety, 60–62, 202n–203n
Beta-blockers, 106
Beta-carbolines, 71, 205n
Biological psychiatry, 93, 209n–210n
Bodily Changes in Pain, Hunger, Fear and Rage (Cannon), 14, 194n

Brain
chemistry of, 63–78
and emotions, 17–22
Brain and Psyche (Winson), 207n
Brave New World (Huxley), 78
Breathing technique, 169–70
The Broken Brain (Andreasen), 210n

C

Cannon, Walter, 14, 75, 194n
Carbon dioxide, 214n
Carotid body, 110
Case histories, 85
Castration anxiety, 43–44
"Catharsis," 199n
Chimpanzees, fears of, 27
Chloride, 205n
Chlorimipramine, 177, 223n
Choices, and anxiety, 8–10
Classical conditioning, 32–35
Classification, 79–94
Cleaning rituals, 174
Cognitive therapy, 108
"Combat neurosis," 181
Community mental health centers, 90–91
Compensation neurosis, 186–88, 225n
Compulsions, 171–73
Computer models, receptors, 77
Concentration camps, 181
Concept of Dread (Kierkegaard), 7, 193n
Conditioned response, 33–34, 37, 46–47
Conditioned stimulus, 33–34
Conditioning, 32–39, 46–47
Conflict
and anxiety definition, 5
and conditioning, 35–36
and neurosis, Pavlov's dog, 35

INDEX 229

Conjectures and Refutations (Popper), 208n
Counterphobias, 44
Counting rituals, 173–74
Crowd phobia, 142

D

DaCosta's syndrome, 113, 213n
Dark woods, fear of, 27
Darkness, fear of, 24, 27
Dead bodies, fear of, 25–26
Death anxiety, 41–42
Deconditioning, 37–39
Defense mechanisms, 5
Delusions, 135–36
Depersonalization, 116–17
Depression
  and agoraphobia, 150, 220n
  and anxiety, 107
  diagnosis, DSM-III, 98–99
  and panic disorder, 214n
  and social class, 119
Derealization, 116–17
Descriptive psychiatry, 93
Diagnosis, 92–94
Diagnostic criteria, 93, 98
Diagnostic and Statistical Manual of Mental Disorders, Third Edition. *See* DSM-III
Disasters, 182, 189
Disease, definition, 80–82, 206n
Displacement, 43, 46
Dive reflex, 110–11
Dread, 7–13
Drives, and conditioning, 33
Drug dependence, benzodiazepines, 61
Drug safety, 60–62
Drug screening, 52–53
Drug tolerance, benzodiazepines, 222n
Drug withdrawal, 59
Drugs. *See* Antianxiety drugs
DSM-III, 92–94
  anxiety disorders, 97–100
  and descriptive psychiatry, 92–94
  influence of, 210n

E

Eating phobias, 140–41
"Effort syndrome," 113
Ego, 40–41
Emotions
  definition, 192n
  and the nervous system, 17–22
Endorphins
  and the brain, 66–70, 204n
  discovery, 70
  review articles, 203n–204n
Envy, 191n
Etorphine, 67
Evolution, 29–30
Exclusion criteria, diagnosis, 98
Existentialism, 8–12
Exposure therapy, 160, 177
Extinction, fear response, 37

F

Facial expression, 7
Familiarization, 37–38
Fear
  versus anxiety, definition, 3–4, 191n–92n
  and conditioning, 32–39
  deconditioning, 37–39
  innate aspects, 23–31
  physiology of, 14–22
Fears and Phobias (Marks), 216n
Flexor reflex, 54–55
"Flooding," 161–62

Flying phobias, 164, 216n
Follow-up studies, 92
The Foundations of Psychoanalysis: A Philosophical Critique (Grünbaum), 206n–207n
"Free-floating anxiety," 5, 45
Freedom, and anxiety, 8–9
Freud, Sigmund
  and American psyche, 86–87
  anxiety definition, 5
  anxiety theory, 40–41, 193n, 200n
  opinion of psychoanalysis, 83–84
  versus J. B. Watson, phobias, 42–47
Frigidity, 141

G

GABA (Gamma-aminobutyric acid), 72, 205n
Gastric response, 22, 196n
Generalization, conditioning, 34
Generalized anxiety disorder, 101–11, 211n
  diagnostic criteria, 101–2
  DSM-III, 97–100
  and "free-floating" anxiety, 5
  panic disorder comparison, 211n–12n, 215n
  treatment, 105–11
Genetic differences, fears, 30
Goldstein, Kurt, 192n
Graded exposure, 158–59
Grünbaum, Adolph, 206n–207n
Guilt, 191n

H

Habits, 33
"Hawk effect," 196n–97n

Heart rate, 20–21, 110, 195n–196n
Heights, fear of, 27–28, 133–34, 217n
Helplessness, 4
Heroin, 67, 203n
Human Nature in the Light of Psychopathology (Goldstein), 192n
Hunt, W. A., 16–17
Huxley, Aldous, 78, 203n
Hyperventilation, 109–10, 169–70
Hypochondriasis, 135, 218n
"Hysteria," 199

I

Id, 40–41
Illness phobias, 134–37
I'm Dancing as Fast as I Can, 59
Imaginary exposure, 158–63
Imipramine, 166–67, 221n–22n
Immunity, and opiate receptors, 69
Implosion therapy, 163
Impotence, 141
Imprinting, 30
"In vivo exposure," 177
Inclusion criteria, diagnosis, 98
Individual variation, 29–31
Inhibition, Symptom and Anxiety (Freud), 200
Innate and learned, distinction, 64
Innate fears, 23–31, 196n, 198n
"Insight" psychotherapy, 108, 223n
Introductory Lectures on Psychoanalysis (Freud), 193n

J

James-Lange theory, 20–21, 75, 195n
Jealousy, 191n
Jones, Ernest, 209n

INDEX

K

Kauffman, Walter, 10, 100
Kierkegaard, Soren, 7–11
Klein, Donald, 220n–22n
Kraepelin, Emil, 93–94

L

Lactic acid, 20, 74, 120. See also Sodium lactate
The Lancet, editorial, psychotherapy, 212n
Landis, Carney, 16–17
"Language of the Body," 17
"Latent" fear, 31
"Law of reciprocal inhibition," 38
Learning. See Conditioning
Learning theory, and phobias, 46–47, 165
Librium
  invention, 51–62
  in phobic anxiety, 163–65
  receptors, 19–20
Lifton, Robert J., 226n
Ligands, 66, 69–70
Little Albert, 45–47, 200n
Little Hans, 43–45, 200n
Locus ceruleus, 75

M

Major depression, 98–99
Major tranquilizers, 57
Malcolm, Janet, 209n
Malingering, 187
MAO-inhibitors
  panic disorder, 121
  phobic disorders, 165–68, 222n
Marijuana, antianxiety effect, 57
Marks, Issac M., 216n, 222n

"Masked" depression, 120
Masturbatory insanity, 80–81
May, Rollo, 41–42, 191n
  anxiety definition, 41
The Meaning of Anxiety (May), 4, 41, 191n
"Medical model," 91
Miltown (meprobamate), 52, 54–55, 58
Mind, Brain, Body (Reiser), 207n
Mitral valve prolapse, 117–19, 214n
Monkeys, fears of, 28–29
Monoamine oxidase inhibitors. See MAO-inhibitors
Morphine, and endorphins, brain, 66–70, 204n
Motor tension, 101
Mutilated bodies, fear of, 25–26

N

Naloxone, 68
Negative reinforcement, 35–36
NeoKraepelinian psychiatry, 93
Nervous system, and emotions, 17–22
Neuron, definition, 65
Neuroses, and conditioning, 34–35
Neurotic anxiety, Freudian definition, 5–6
Neurotransmitters, 65
New York Review of Books (Malcolm), 209n
The New Yorker (Malcolm), 209n
Norepinephrine, 75
Nostalgia, 80, 179, 205n
Novelty, fear of, 28–29

O

Obsessive compulsive disorder, 171–72, 223n

Obsessive compulsive disorder (cont.)
diagnostic criteria, 171–72
DSM-III, 97–100
forms of, 172–73
versus obsessive personality, 175–76
treatment, 176–77, 223n
Obsessive personality, 175–76
One-trial learning, 199n
Operant conditioning, 35–36, 201n
Opiate antagonists, 68–69
Opiate receptors, 67–70
Opiates, and endorphins, brain, 66–70
Opium, 66–67
Overbreathing, 109–10, 169–70

P

Pain, and opiate receptors, 69
Panic disorder, 112–22, 213n
and agoraphobia, treatment, 166–67
cause, 119–20, 213n–14n
classification, 211n
definition, 192n
and depression, 120, 214n
diagnostic criteria, 113–14
DSM-III, 97, 113–14
generalized anxiety disorder comparison, 211n–12n, 215n
lactic acid, 74–75
PET scan, 205n
social class, 119
"Paradoxical intention," 109
Paraplegics, 20, 195n
Parasympathetic nervous system, 18–19
"Participant modeling," 160
Pascal, 8
Pavlov, Ivan P., 32–35
Pavlov's dog, 32, 34–35

Penis envy, 44
Performance fear, 138–39
Phenelzine, 165–67
Phenobarbital, 57
Phobic disorders. *See also* Agoraphobia; Simple phobia; Social phobia
alcoholism, 216n
behavior therapy, 158–63
book recommendations, 221n
drug treatment, 163–70, 222n
DSM-III, 97–100
Freudian interpretation, 5
history, 126–27
natural history, 136–37
prevalence, 125
prognosis, 137
simple type, 128–37
social type, 138–44
treatment, 157–70
Watson versus Freud, 42–47
Phobos, 126
Popper, Karl, 84–85, 200n, 208n
Positive reinforcement, 35–36
Positron emission tomography, 72–74, 205n
Post-traumatic stress disorder, 178–.89, 224n
delayed type, 186–88
diagnostic criteria, 183
DSM-III, 97, 183
historical background, 178–83
treatment, 188–89
Post-traumatic stress disorder, delayed type, 186–88
Premature ejaculation, 141
The Problem of Anxiety (Freud), 193n, 200n
Progressive relaxation, 159, 222n
Projection, *Little Hans*, 43
Propranolol, 106, 122
in phobias, 168–70

Psychiatric Diagnosis (Goodwin and Guze), 223n
Psychoanalysis, 82–89, 167, 206n–209n
 and "substitute" symptoms, 167
Psychopharmacology, 90
Psychotherapy
 evaluation of, 107–9, 212n–13n
 in phobias, failure, 157
Psychotherapy by Reciprocal Inhibition (Wolpe), 157
Public speaking, fear of, 139

Q

Quadraplegics, 20, 195n

R

Reason, 11–12
"Rebound" anxiety, 59
Receptors, 19–20, 65–66
 computer models, 77
 definition, 65–66
 and drugs, 76–78
 endorphins, brain, 66–70
 opiates, 67–68
Reciprocal inhibition, 38, 159, 222n
Reinforcement, 35–36
Reiser, Morton, F., 207n
Relaxation therapy, 159, 222n
"Releasing mechanisms," 24
Reliability, 82
Religious faith, 194n
"Repetition compulsion," 180
Repression
 and anxiety definition, 5
 Little Hans, 43–44
Reward and phobias, 46
Righting reflex, 54–55
Rituals, obsessional, 173–74, 223n
RO5-0690, 54–55
Roche Laboratories, 51–56

S

Sartre, Jean Paul, 9
Scanning, 102
Schizophrenia, DSM-III, 98
School phobia. *See* Separation anxiety
"Secondary" depression, 106–7
Sedatives, 57
Separation anxiety, 152–56, 221n
 diagnostic criteria, 152–53
 imipramine in, 167, 221n
Sexual performance, and phobia, 141
Shame, 191n
Shell shock, 179, 224n
Simple phobias, 128–37
 diagnostic criteria, 128–29
 drug treatment, 164
Skeletal nervous system, 18
Snakes, fear of, 26–27
Social class, and depression, 119
Social phobia, 138–44
 alcoholism, 216n
 diagnostic criteria, 138
 drug treatment, 164
 prevalence, 125
Society for Neuroscience, 63
Sodium lactate, 120, 205n, 214n.
 See also Lactic acid
Soteria, 38
Spitzer, Robert L., 92–93
Stage fright, 139, 168–69
Staring eyes, fear of, 28, 218n
The Startle Pattern (Landis and Hunt), 194n

Startle reaction
  and fear, 16–17
  and trauma, 184
Starwave (Wolf), 207n
"State-dependent learning," 167
Sternbach, Leo, 51–55, 201n
Stimulus characteristics, and fear, 30–31
Stomach, anxiety response, 22, 196n
Strangers, fear of, 24–25, 197n
Subconscious. *See* Unconscious
Sublimation, 5
"Substitute" symptoms, 167
Summation, 31
Superego, 40–41
Supportive psychotherapy, 107
  phobic disorders, 166–67, 215n
Symbolization, *Little Hans*, 43–44, 46
Sympathetic nervous system, 18–19
Symptom clusters, 81
Synapse, 65. *See also* Receptors
Syphilophobia, 217n–18n
Systematic desensitization, 159

T

Temperament, 198
Thorazine-type drugs, 57, 89–90, 163
Thought-editing, 108–9
Time magazine article, Thorazine, 89
Toilet phobias, 139–40
Touching, fear of, 143
Tranquilizers. *See also* Benzodiazepines
  invention, 51–62
  in phobic anxiety, 163–65
Trauma, 178–89, 224n–25n
Tricyclic antidepressants, 121. *See also* Imipramine
panic disorder, 121, 166
phobic disorders, 165–68

U

U-shaped model, 21
Unconditioned response, 33–34, 46–47
Unconditioned stimulus, 33–34
Unconscious
  and anxiety definition, 5
  phobias, 43–44
  Walter Cannon's criticism, 14–16
Ungraded exposure, 161–62

V

Validity, 82
Valium
  invention, 56–62
  in phobic anxiety, 163–65
  receptors, 19–20, 70–72
Vietnam War, 181–89, 224n–25n
Vigilance, 102

W

Washington University in St. Louis, 91–93
Watson, John B., 40–47
  versus Freud, phobias, 42–47
Westphal, Karl, 126, 145, 149, 216n
Winson, Jonathan, 207n
Withdrawal symptoms, 59, 61, 222n
Wolf, Fred Alan, 207n
World Health Organization, 61–62

X

Xanax (alprazolam), 121, 164

Help is Here!

A Spiritual Survival Manual
for Times of Crisis.

C.J. DeLong

Help is Here!

Copyright © 2002
All rights reserved.

Library of Congress Control Number: 2001088932

This book, or any portion, may not be reprinted
by any means without permission.

Printed in the United States of America

First Edition

Published by
Excelsior House, Inc.

For information:
www.ExcelsiorHouse.com
info@excelsiorhouse.com

ISBN 0-9706303-1-X / ISBN 0-9706303-0-1 *(Alternate Cover)*

EXCELSIOR HOUSE PUBLISHING

Dedication

To my family and friends:

For your patience and understanding.
For the gift of laughter.
For your help, and knowing when not to help.
For all of the little acts of kindness and friendship.
For your thoughts and prayers.
For your inspiration.
For your love.
Thank You.

Special thanks
to my parents, whose unwavering love and support
has provided a foundation for my faith.
And to the others I leaned on, whose love and
friendship grew only stronger with the weight.

Above all, I thank God, who has filled me with the
Holy Spirit, through Jesus Christ, and given me the gifts of
peace, guidance, and healing.

ACKNOWLEDGEMENTS

I wish to thank everyone whose words of inspiration, comfort and encouragement helped me along the way. My deepest gratitude to the following people:

All of the authors, known and anonymous, whose writing is included in this book.

Lecy Design, for their kind help (and great patience) in pulling the book together.

My dear friend Lynn, for her endless encouragement, and for inspiring me, by shining example, to grow in faith and purpose.

Van and Molly, for their enthusiasm and help with production — and for sharing my niece and nephew who make it so easy to find joy.

My mother, for her beautiful painting which was the answer to my cover design; like her it has always been a wonderful source of comfort and peace.

My dad, for sharing his quiet devotion to scripture. The strength of his faith is forever etched in my soul.

Used by permission:

Pg 6	*Who Needs God*, Harold Kushner. Simon & Schuster, New York, NY, 1989.
Pg 14, 62	*Psalms/Now*, Leslie Brandt, Concordia Publishing House, St Louis, Missouri, 1973.
Pg 20, 73	*Daily Word*, Colleen Zuck. Rodale Press, Emmaus, Pennsylvania, 1997.
Pg 28	*100 Ways to Keep your Soul Alive*, Frederic & Mary Ann Brussat. Harper, San Francisco, 1994.
Pg 60	*The Will of God*, Leslie D Weatherhead. Abingdon Press, Nashville, Tennessee, 1972.

I would also like to acknowledge Stanford for its course in creativity and meditation, and the Tai Chi class at The Marsh for helping me learn to breathe again.

Contents

First Call For Help - 1

Know That You Are Not Alone - 5

Gain The Strength To Heal - 19

Don't Worry About Tomorrow - 39

Find Meaning And Purpose - 57

*Then you will call
and the Lord will answer;
you will cry for help
and he will say Here Am I.*

Isaiah 58:9

First Call for Help

If I can stop one heart from breaking,

I shall not live in vain:

If I can ease one life the aching,

Or cool one pain,

Or help one fainting robin

Unto his nest again,

I shall not live in vain.

~Emily Dickenson

First Call For Help

It has occurred to me, looking back on it, that my immediate reaction to bad news (shocking, devastating news) is to repeatedly say "Oh My God, Oh My God". The interesting thing is that without any conscious thought at all, instinctively my first call for help is to God. It seems only natural then that I should follow by finding comfort and strength and courage in the words of the Lord — be it through psalms, poetry, prayer, or other reference.

I first put together this book for my own use while battling a life threatening illness. I found myself writing poetry as though it flowed from an unknown source. I logged miscellaneous journal entries. I wrote a series of meditative and faith strengthening "exercises". I also wanted one place where I could collect and refer back to the various readings that helped me get through many a sleepless night; words of inspiration that had come to me by way of friends and family, and often seemed to appear out of nowhere just when I needed them most. My father handed me his prayer book; my mother her book of psalms. A friend sang a hymn for me on NPR one night and I was reminded of the power of their lyrics.

I hope now to share these thoughts with anyone else who could use a little peace of mind.

C.J.D.

*Each morning lean your arms awhile
upon the windowsill of heaven
and gaze upon the Glory of the Lord.*

*Then with the vision still fresh in your heart
go forth strong to meet the day!*

Know *that* You *are* Not Alone

Turning to God

RELIGION BEGINS WITH A SENSE OF REVERENCE, THE RECOGNITION OF GOD'S GREATNESS AND OUR LIMITATIONS. That is why there are no atheists in foxholes and few atheists in hospitals. It is not because people are hypocrites, ignoring God when things are going smoothly and suddenly discovering Him and pleading piety when they are in trouble. And it is not just a matter of turning to God out of fear. There are no atheists in foxholes because times like those bring us face to face with our limitations. We who are usually so self-confident, so secure in our ability to control things, suddenly learn that the things that matter most in our lives are beyond our control. At the limits of our own power, we need to turn to a Power greater than ourselves.

Because God has so much strength and power, He can supply us with the strength we need when we face challenges that exceed our human capacities.

Harold Kushner, Who Needs God

From the author's journal...

THE PEACE OF HIS PRESENCE

A couple of weeks before my health crisis arose I began to wish I had a cross for the chain I wear everyday around my neck. Two weeks later, on the way to visit the doctor, I had some extra time and stopped in a store to distract myself. My eyes fell on a beautiful cross, which I purchased and immediately hung on the chain I was wearing. The next day I was sent to the hospital for a series of tests. Frightened to the point of shaking, with difficulty breathing, my panic continued on through various technicians' work. I was unable to calm myself. At one point everyone left the room as I waited for the test results. Remembering that although I'd removed the chain from my neck it was in my pocket, I reached for it, thinking I would say one more prayer. To my surprise, as my fingers touched the cross I felt the most amazing sense of peace. My panic was replaced by a relaxation so deep I nearly fell asleep! I was filled with the Holy Spirit and knew that my life had changed forever. The absolute knowledge of God's presence within me was a tremendous source of comfort – not only at that frightening moment but throughout all of the difficult times ahead.

"THE PEACE WHICH PASSETH ALL UNDERSTANDING SHALL BE YOURS THROUGH JESUS CHRIST"
PHILLIPIANS 4:7

ON EAGLE'S WINGS
(HYMN)

You who dwell in the shelter of the Lord
Who abide in this shadow for life
Say to the Lord "My refuge...my rock
in whom I trust!"

And I will raise you up on eagle's wings
Bear you on the breath of dawn
Make you to shine like the sun
And hold you in the palm of my hand

Snares of the fowler will never capture you
and famine will bring you no fear
Under God's wings your refuge
with faithfulness your shield

For to the angels God's given a command
To guard you in all of your ways
Upon their hands they will bear you up
Lest you dash your foot against a stone

And I will raise you up on eagle's wings
Bear you on the breath of dawn
Make you to shine like the sun
And hold you in the palm of my hand

TEXT AND TUNE: MICHAEL JONEAS 1979

"Have you not known? Have you not heard?
The Lord is the everlasting God, the Creator of the ends
of the earth. He does not faint or grow weary;
His understanding is unsearchable.
He gives power to the faint,
and strengthens the powerless.
Even youths will faint and be weary,
and the young will fall exhausted, but those who wait for
the Lord shall renew their strength,
they shall mount up with wings like eagle,
they shall run and not be weary,
they shall walk and not faint."

Isaiah 40:28-31

COURAGE IS NOT SIMPLY ONE OF THE VIRTUES, BUT THE FORM
OF EVERY VIRTUE AT THE TESTING POINT.

C.S. Lewis

FOOTPRINTS

ONE NIGHT A MAN HAD A DREAM. He dreamt he was walking along the beach with the Lord. Across the sky flashed scenes from his life. For each scene he noticed two sets of footprints on the sand – one belonging to him and the other to the Lord. When the last scene had flashed before him, he looked back at the footprints and he noticed that many times along the path there was only one set. He also noticed that this happened during the lowest and saddest times of his life. This bothered him and he questioned the Lord. "Lord, you said that once I decided to follow you, you would walk all the way with me, but I noticed that during the most troublesome times of my life there was only one set of footprints. I dont understand why when I needed you most, you deserted me." The Lord replied, "My precious child, I love you and would never leave you. During your times of trial and suffering, when you see only one set of footprints, those were the times when I carried you in my arms."

ANONYMOUS

I lift up mine eyes to the hills, from whence commeth my help.

Help comes only from the Lord, maker of heaven and earth.

How could He let your foot stumble? How could He your guardian sleep?

The guardian of Israel never slumbers, never sleeps.

The Lord is your guardian, your defense at your right hand;

The sun will not strike you by day nor the moon by night.

The Lord will guard you against all evil;

He will guard you body and soul.

The Lord will guard your going and your coming now and for evermore.

PSALM 121

From the author's journal...

IT'S IN YOUR HEART

The first night in the hospital after surgery I panicked for a moment when I realized my cross was missing. I'd removed the chain with my cross prior to surgery and now had no idea where it was. I buzzed for a nurse's assistant and a nice young man with a very strong accent came to help. I asked him to look around for my cross. After a minute or so he found it and held it up near me but just out of reach. He said something I didn't quite catch and he had to repeat it several times. Finally I understood: he said "It's in your heart you know."

"I will ask the Father, and he will give you another Counselor to be with you forever — the Spirit of truth. ... you know him, for he lives with you. I will not leave you as orphans; I will come to you ... Because I live you also will live ... The Holy Spirit whom the Father will send in my name, will teach you all things and will remind you of everything I have told you. Peace I leave with you; my peace I give you. I do not give to you as the world gives. Do not let your hearts be troubled and do not be afraid."

JOHN 14:15-28

"I pray that out of His glorious riches
He may strengthen you with the power through His Spirit
in your inner being, so that Christ may dwell in your heart
through faith. And I pray that you being rooted and established
in love, may have the power with all the saints to grasp
how wide and long and high and deep is the love
of Christ, and to know this love that surpasses
all knowledge — that you may be filled to the
measure of all the fullness of God."

EPHESIANS 4:16-19

My Heart Is Secured

How great is my God and how I love to sing His praises!

Whereas I am often frightened when I think about the future, and confused and disturbed by the rapidly changing events about me, my heart is secured and made glad when I remember how He has cared for me throughout the past.

When I was brought forth from my mother's womb, God's hand was upon me. Through parents and people who cared, He loved and sheltered me and set me upon this course for my life.

Through illness and accident my God has sustained me. Around pitfalls and precipices He has safely led me.

When I became rebellious and struck out on my own He waited patiently for me to return.

When I fell on my face in weakness and failure He gently set me on my feet again.

He did not always prevent me from hurting myself, but He took me back to heal my wounds.

Even out of the broken pieces of my defeats He created a vessel of beauty and usefulness.

Through trials and errors, failures and successes, my God has cared for me.

From infancy to adulthood He has never let me go.

His love has led me – or followed me – through the valleys of sorrow and the highlands of joy, through times of want and years of abundance.

He has bridged impassable rivers and moved impossible mountains.

Sometimes through me, sometimes in spite of me, He seeks to accomplish His purposes in my life.

He has kept me through the stormy past, He will secure and guide me through the perilous future.

I need never be afraid, no matter how uncertain the months or years ahead of me.

How great is my God, and how I love to sing His praises.

Leslie Brandt, Psalms/Now, 105

In thee O Lord I have taken refuge; never let me be put to shame.[1]

As thou art righteous rescue me and save my life;
hear me and set me free,[2]

Be a rock of refuge for me, where I may ever find safety at thy call; for thou art my stronghold[3]

O God, keep my life safe from harm[4]

Thou art my hope O Lord, my trust O Lord[5]

From birth I have leaned on thee,
my protector since I left my mother's womb.[6]

My mouth shall be full of thy praises,
I shall tell of thy splendor all day long.[8]

Do not cast me off when old age comes,
nor forsake me when my strength fails.[9]

O God do not stand far from me; hasten to my help.[12]

I will wait in continual hope, I will praise thee again and again[14]

Thou hast made me pass through bitter and deep distress, yet revives me once again.[20]

Restore me to honor, comfort me.[21]

All day long I shall tell of your righteousness;
shame and disgrace await those who seek my hurt.[24]

PSALM 71

With His Help I Am...

The one I turn to
when I'm down,
who lifts me up, says carry on.
Who judges with kindness,
a voice of intelligence.
Who when I don't know,
answers yes or no.
The one who's always there,
who always cares,
who watches out,
who questions why,
observes my day,
how I behave,
who takes my fears away.
Who wakes me with a smile
and says that it's okay.

Who knows no one but me
and therefore can't compare.
Who's walked a mile
in my shoes
and not tired of the ride,
says be brave and just do
for whom everything is new.
So I will
be good to myself,
take care of myself,
have faith
and trust in myself ...
With His Help.

C.J.D.

Experience has proven that defeat can be turned into victory when we have the necessary dedication. The strength to go on struggling and salvaging what can be salvaged has saved many lives from disaster.

GAIN *the* STRENGTH *to* HEAL

Healing Energy

"Then your light shall break forth like the dawn,
and your healing shall spring up quickly."
Isaiah 58:8

As I close my eyes I begin to visualize myself aglow with life. God's healing energy is in every atom and cell of my body. I am immersed in healing, purifying light.

If there is an area of special need, I focus my attention on it and know that God's healing energy is doing its work there.

I let go of any concern and trust the divine activity within that heals me and keeps me well and whole. I relax and let God's healing light renew me in mind and body.

God guides me on the path of health and healing, and I visualize this same healing energy working within my loved ones. No matter what the need, God's healing energy is greater.

In God's loving care, I am charged with new life.

Daily Word, Edited by Colleen Zuck

Life is 10% what happens to you,
90% what you do about it.

From the author's journal...

EMOTIONAL & PHYSICAL HEALING

When I was first told that I had a life threatening illness, one of the things the doctor said to me was "you need to have a positive attitude because it really does matter". I remember wondering at that moment how I would ever get from my current state, which was one of unbelievable shock and horror – a sort of terrified numbness – to anything remotely resembling a positive attitude. Trying to explain to someone how I felt left me searching for words; perhaps my aunt put it best, "you feel like you want to curl up on the closet floor and suck your thumb." How do you get from there to a place of strength from which to heal?

I found there are deliberate ways to begin healing emotionally. Whether the crisis is a physical or emotional trauma, the two are rarely entirely separable, and the healing of one generally goes hand in hand with the other. Things like meditation and visualization can directly impact the physical healing process by untapping the great power of the mind. Feeding your soul with family, friends and laughter can comfort and literally soothe raw nerve endings. Focusing on the simple pleasures in life can bring on a smile in the direst of circumstances (and scientists have proven that the physical act of smiling creates a "feel good" response in the brain). A determined faith in God through word and prayer can work miracles of both mind and body.

An Exercise of Faith

1. Say these words: "BE STILL, AND KNOW THAT I AM GOD" (Psalm 46:10). Then close your eyes, be still and relax physically.

2. Say these words: "THOU DOST KEEP HIM IN PERFECT PEACE, WHOSE MIND IS STAYED ON THEE" (Isaiah 26:3). Then relax mentally. Fear and lack of confidence come from my placing obstructions in the channel through which God's spirit reaches me. The greatest obstruction is self.

3. Say these words: "IF YOU ABIDE IN ME... ASK WHATEVER YOU WILL, AND IT SHALL BE DONE FOR YOU" (John 15:7). Then rest in Him spiritually. Remember "He is able to help" (Hebrews 2:18). "He is able for all time to save those who draw near to God through Him" (Hebrews 7:25). "Him who is able to keep you from falling" (Jude 24). Not I but He is able, He is able.

4. Say these words: "IN QUIETNESS AND IN TRUST SHALL BE YOUR STRENGTH" (Isaiah 30:15). Then breathe deeply and evenly as in sleep. Picture the Holy Spirit "creating a little pool of heavenly peace" within your heart. Remember Tennyson's words: "Prayer is like opening a sluice between the great ocean and our little channels when the great sea gathers itself together and flows in at full tide."

5. Say these words: "KNOW THAT YOUR BODY IS A TEMPLE OF THE HOLY SPIRIT WITHIN YOU" (I Corinthians 6:19). Say "O Living Christ, I am conscious now of thy healing nearness. Thou hast touched my eyes and I see Thee; Thou hast opened my ears and I hear Thy voice; Thou hast entered my heart and I have Thy love. Thou dost over shadow my soul and body with Thy presence, therefore I am filled with Thy strength, Thy love, and Thy healing life."

6. Repeat over and over again thoughtfully the thing you want God to accomplish in your life: "HEAL ME, O LORD, AND I SHALL BE HEALED" (Jeremiah 17:14a).

7. Then, because the prayer of faith at once sets in motion God's healing power; give thanks to God through Jesus Christ. Amen.

On Wings of Healing

"FAITH IS BEING SURE OF WHAT WE HOPE FOR
AND CERTAIN OF WHAT WE DO NOT SEE."
Hebrews 11:1

The First Ray

catching my eye,

I stand silently,

watching the sun rise —

slowly over the morning horizon,

warming the dew covered land

and separating day

from the dark of night.

C.J.D.

Visualization

Use the power of your mind through focus and concentration to directly influence and mobilize the physical healing process:

First, relax physically. Focus on your breathing and relax each muscle group.

Next, visualize a pleasant, natural scene such as the beach at sunset. Watch the waves moving in and out, feel the warmth of the sun and the sand at your feet, hear the sounds of the ocean and the seagulls, feel the breeze.

Now, visualize your affliction in some physical way and begin to visualize whatever your form of treatment acting upon the affliction. You might, for example, choose to see your illness as a threatening animal, then see your treatment attacking it; leaving it unable to do further damage, subdued and transformed. Or you may want to visualize cells being acted upon by chemicals designed to destroy them; poisoning the bad cells and leaving good cells to thrive intact.

End by visualizing the growth of your favorite tree as water nourishes the seed to root and grow a trunk, branches and leaves. Look all around you as you expand your energy in all directions. Be filled with healing life.

Quietness

"In quietness and in trust shall be your strength"
Isaiah 30:15

Sit comfortably. Gently close your eyes and relax. Breathe easily and deeply from center. Go into silence. Notice the rise and fall of the abdominal breathing. Look inside. Notice the sound of the inside. Experience the quiet inside. All that is of value is there. All of your useful thoughts, emotions, and insights are inside. Your body is a container for them. Feel inside as quiet as you can feel. If negative thoughts come, just let them go. Think of them as clouds passing by. You and your mind can relax. Now go even deeper and feel ever more quiet. Imagine yourself as an empty space with a light inside you. Feel suffused with this light. Now become that light. Feel the energy and strength of that light. When you are ready slowly open your eyes and your ears.

As you go about your daily activity try simply repeating a sound or a few words in your mind to block out the useless chattering and turn away all worry and negative thoughts. For example, repeat: "Don't Worry, Breathe Deeply".

Research at the Mayo clinic suggests that a positive attitude results in a 20% improvement in recovery.

Smile

Be open to the slightest opportunity for happiness; then keep that feeling with you as much as possible:

> Play your favorite CD.
> Listen to your favorite hymn.
> Play an instrument if you can.
> Sing in the choir, or the shower.
> Spend time with a child.
> Rent a funny movie.
> Read a funny book.
> Watch classic sitcoms on tv.
> Cook your favorite meal.
> Eat ice cream.
> Light candles.
> Sit by the fire.
> Make snow angels.
> Blow bubbles.
> Buy yourself flowers.
> Smell the roses.
> Go for a walk in the park.
> Watch the sunset.
> Smile.

LAUGHTER IS A TRANQUILIZER WITHOUT SIDE EFFECTS.

Feed Your Soul

"I AM TRYING TO KEEP MY SOUL ALIVE IN TIMES
NOT HOSPITABLE TO SOUL."
J.M. COETZEE, *AGE OF IRON*

KEEPING YOUR SOUL ALIVE BECOMES URGENT DURING TOUGH TIMES IN YOUR LIFE. Here are a few things to recognize in the effort to heal the soul:

- The soul savors the present moment. You don't find it hiding out in the past or waiting for you in the future. It says pay attention to what is happening to you right now.

- The soul speaks its own peculiar language in the messes and miseries of life. It does not run from trouble or lift itself above the fray. Don't think you have to fix everything about yourself for your soul's sake.

- The soul savors simplicities — love that is true, wonder that is childlike, humility that is homegrown.

- The soul is always on the lookout for fresh wonders. It likes taking the long way home. Be patient with its haphazard and zigzag meanderings.

- *The soul reaches out to others through love, compassion and forgiveness. Let its music play through your words and deeds.*

- *The soul satisfies itself and finds fulfillment in community. Recognize that your family, friends, neighbors, and colleagues all represent opportunities for spiritual collaboration.*

- *The soul's embrace of people and things is wide and welcoming. It is hospitable to strangers and revels in the opportunity to be an angel for another soul in need. Never constrain this impulse.*

- *The soul seeks out silence and solitude in order to hear the soft voice of God. Treasure the quiet times and make a place for them in your busy life.*

- *The soul is fed by lifelong learning. Many sacred texts and other resources will speak to your deepest needs and illuminate your path. Remember that a little study each day is good for the soul.*

FREDERIC & MARY ANN BRUSSAT, *100 WAYS TO KEEP YOUR SOUL ALIVE*

FAMILY, FRIENDS, AND A SENSE OF HUMOR
WILL SERVE YOU WELL.

From the author's journal...

PRAY

I couldn't begin to count the number of people who have prayed for my recovery. It's like an ever-widening circle: friends and family, friends of friends and family, people at work, people at church ... and it means everything. I have found that the two most comforting things you can hear from someone are "you will be alright" and "you are in our prayers." The first is true because the second is done.

Why pray? Because we are human, we need help. Prayer gives us the courage and the strength through the power of God to do what needs to be done. Prayer can lead us to a clearer vision of who we are and what we are capable of. If we pray for guidance, peace, and healing, we will receive these gifts. Ask and He will answer. But we must also learn to listen carefully — be still and engaged in thought. Prayer is an amazing force which must be experienced to be believed. I never cease to express my amazement at the miracles, big and small, I have experienced and witnessed. And yet, I shouldn't be surprised, they are ours to expect. "Ask and it shall be given to you."

"REJOICE IN YOUR HOPE, BE PATIENT IN TRIBULATION, BE CONSTANT IN PRAYER."
ROMANS 12:12

WHAT A FRIEND WE HAVE IN JESUS
(HYMN)

What a friend we have in Jesus. All our sins and griefs to bear.
What a privilege to carry everything to God in prayer.
Oh what peace we often forfeit. Oh what needless pain we bear.
All because we do not carry, everything to God in prayer.

Have we trials and temptations? Is there trouble anywhere?
We should never be discouraged. Take it to the Lord in prayer.
Can we find a friend so faithful, who will all our sorrows share?
Jesus knows our every weakness, take it to the Lord in prayer.

Are we weak and heavy laden, cumbered with a load of care?
Precious Savior still our refuge, take it to the Lord in prayer.
Do your friends despise, forsake you?
Take it to the Lord in prayer.
In His arms He'll take and shield you,
you will find a solace there.

TEXT: JOSEPH SCRIVEN 1820-1886
TUNE: CHARLES CROZAT CONVERSE 1832-1918

Prayers

Evening Prayer

Thank you Lord for helping me to know your nearness to me now and at all times. Please watch over me every day, in every way. Keep me safe from harm and heal me. Thank you for blessing me with family and friends and their loving support. Help me to meet tomorrow with confidence, courage, strength and peace. Grant me a full and long life and let my thanks be in service to you. Amen. C.J.D.

Morning Prayer

Lord, help me to take each day as it comes, to live in the present moment; this is the day which you have made — let me rejoice and be glad in it. Don't allow to me spend my time today worrying about tomorrow or regretting yesterday. Tears are precious, don't let me waste them. Lord, let me know that you are forever with me and that I alone would suffer from fate, but together with you I can handle anything. Amen. C.J.D.

Prayer For a Cheerful Outlook

Keep me, O God, from all complaining and self-pity. When my work is hard, give me the strength I need to do my best. When my heart is heavy, give me a sense of Thy many blessings and now and always may my efforts and my hopes make me cheerful and serene, that others may be inspired with new vigor and joy. Amen. Anonymous

Prayer For Healing

O God of all healing ease my pain. Through the daylight hours keep me confident and cheerful. All through the night give me rest and refreshing sleep. Hasten my recovery and help me in patience and courage to wait for the day when, restored in health and strength, I go again to the tasks which lie before me. Amen.

My Book of Prayers,
Hammerberg & Nelson

Prayer Before Surgery

Into your fatherly keeping, O Lord, I commit my life completely with trust and confidence. Thou art my dwelling place, and underneath are your everlasting arms. To you I give my body that you may repair it. May I fall asleep peacefully in Thee. Guard me through the moments of unconsciousness, guide the hands and give skill to the surgeon that through the wounds he must make there might enter in your healing power to restore me to health and strength. Cheer and sustain my family and my loved ones who stand by and wait. I ask this, O Lord, in the name of Jesus Christ. Amen.

My Book of Prayers,
Hammerberg & Nelson

Bless the Lord, O my soul:

and all that is within me,

bless His holy name.

Bless the Lord, O my soul,

and forget not all His benefits.

Who forgives all thine iniquities,

who healeth all diseases,

Who redeemeth thy life from destruction,

who crowneth thee with loving kindness

and tender mercies,

Who satisfieth thy mouth with good things,

so that thy youth is renewed like the eagles.

PSALM 103:1-5

From the author's journal...

READ THE BIBLE

I was in the middle of a good book around the time my crisis began. When I first picked up the book to distract from my distress the bookmark fell open at the paragraph below. Prior to that point the book contained no religious reference whatsoever. The timely reminder to reach for the Bible as a source of tranquility led me to Psalms and other passages of immense comfort. I've since added a little reading to my morning routine and found it to be a remarkably simple source of daily strength.

"Now that they had won each other's confidence she learned that he read in the Bible for an hour upon awakening and also before going to sleep at night. Perhaps that is the source of his tranquility she thought; anyone who grows up hand in hand with death must have a feeling of intimacy with God. When she listened to him speak with reverence of God's will she knew that he alone had inherited his fathers religious nature; and that seemed doubly wondrous to her, for he not to have inherited his fathers eyes or chin or voice or wilderness fever but precisely the one thing he needed to sustain him: devoutness. In her confused and frightened state of mind she had found a brother whose very nature fitted her."

IRVING STONE, THE PRESIDENT'S LADY

BECOME CENTERED AND BALANCED

Find a class and try one of the ancient forms of revitalization and relaxation, like gentle Yoga or Tai Chi, to become more centered and balanced.

If I continually withdraw in fear,
I am tipping backward, tense and rigid
And the slightest surprise will push me over.

If I feel uncertain in myself,
and unstable in my base,
Then all my contacts with others
will be wobbly and lack conviction.

If I can become centered and balanced
In my own experience,
Then I can carry this moving center with me.

If I am balanced now…
Then I can move in any direction I wish,
with no danger of falling.

AL CHUNG-LIANG HUANG

"SURVIVORS OF GREAT LOSS OR TRAUMA OFTEN RADIATE A DEEPLY GROUNDED PEACE, NOT BECAUSE THEY'VE NEVER LOST THEIR CENTER, BUT BECAUSE THEY'VE BEEN KNOCKED INTO WHOLE DIFFERENT GALAXIES, ONLY TO REALIZE THEY COULD STILL FIND THEIR WAY BACK TO THEMSELVES."

MARTHA BECK

Reflect In The Beauty Of Nature

If possible, retreat into nature; even if only for a day.

In nature there is the opportunity,

in solitude and stillness, to experience

the true abandonment of your cares and worries.

To focus on nothing more than a blade of grass.

To echo the cry of a loon.

To submerge yourself in cool waters.

To make what you will of the clouds floating by.

To hear the birds welcome the day in song.

To note the passing of time only as the sun sets

in the sky. To know that life itself is a miracle.

C.J.D.

Leave the future to God's providence,
Leave the past to God's mercy,
And live in the present moment.

Don't Worry about Tomorrow

Some of your hurts you have cured,

and the sharpest you still have survived.

But what torments of grief you endured

From the evils that never arrived.

~ Ralph Waldo Emerson

From the author's journal...

FASTING FROM WORRY

In the face of dire circumstances, focusing on positive thought takes a great deal of energy. But think of all the energy it takes to worry (staying awake all night, conjuring up ways to continuously replace one worry with another, sustaining some measure of worry even in the best of times). I finally decided to transfer some of that energy toward prayer and meditation. In doing so, I began to block the negative thoughts and fears from my mind and find the peace that comes from living in the moment.

The one thing sure to stand between us and God's healing power is a negative, closed mind. To keep your mind open: 1) be focused in the present, and 2) compartmentalize the fearful experience. After all, it is only one aspect of your life, and you must deal with it, but it is not the essence of you or your life. Think of it, act on it, when necessary, at specific times, then set it aside and free yourself to live the many other facets of your life whenever possible. A friend of mine decided to put this into practice during Lent, and joyfully reported back on the time she spent "fasting from worry."

Prayer on Worry

Father, teach me that, as your child, worry has no place in my life. I know that it helps nothing. I know that by worrying I cannot add a single cubit to my stature. I know that fretting overcomes no difficulty. Often in the past, Lord, I have come to you with a heavy heart and burdened life. And you have answered my prayers and graciously lifted the burden from me. Yet with a strange perversion, I still refuse to leave my burdens with you. Always I gather them up — those heavy bundles of fears and anxieties — and shoulder them again. Do for me what I cannot do for myself. Break these habit patterns, reverse the direction of my negative thoughts, lift from me once again all anxieties and apprehensions. Give me in their stead a calm and confident trust in you. Make me willing to live just one day at a time. May my heart re-echo to your promise that only as I rest in you can the desires of my heart be given me. And now help me to do my part by placing a guard around my thoughts, by resolutely refusing to return to my old haunts of distrust. I thank you for your love for me and for your help. Amen.

My Book of Prayers, Hammerberg & Nelson

If you're worrying,
you're not praying.

Yesterday, Today & Tomorrow

There are two days in every week we should not worry about, two days that should be kept free from fear and apprehension.

One is yesterday, with its aches and pains. Yesterday has passed, forever beyond our control. All the money in the world cannot bring back yesterday. We cannot undo a single act we performed. Nor can we erase a single word we've said ... yesterday is gone.

The other day we shouldn't worry about is tomorrow, with its impossible adversaries, its burden, its hopeful promise and poor performance. Tomorrow is beyond our control. Tomorrow's sun will rise either in splendor or behind a mask of clouds, but it will rise. And until it does we have no stake in tomorrow, for it is yet unknown.

This leaves only one day: Today. Any person can fight the battles of just one day. It is only when we add the burdens of yesterday and tomorrow that we break down. It is not the experience of today that drives people mad; it's the remorse of bitterness for something that happened yesterday, and the dread of what tomorrow may bring.

Anonymous

Worry does not empty tomorrow of its troubles;
it does empty today of its strength.

Therefore I tell you, do not worry about your life.

Look at the birds of the air; they do not sow or reap or store away in barns, and yet your heavenly Father feeds them.

Are you not much more valuable than they?

Who of you by worrying can add a single hour to his life?

See how the lilies of the field grow. They do not labor or spin.

Yet I tell you that not even Solomon in all his splendor was dressed like one of these.

If that is how God clothes the grass of the field, which is here today and tomorrow is thrown into the fire, will He not much more clothe you, O you of little faith?

So do not worry.

But seek first His kingdom and His righteousness, and all these things will be given to you as well. Therefore, do not worry about tomorrow, for tomorrow will worry about itself. Each day has enough trouble of its own.

MATTHEW 6:25-34

Don't Worry, Breathe Deeply

Are there moments when you realize you have gone great lengths of time without seeming to breathe at all? Taking time out to do the following exercise not only reminds the body to breathe, but establishes a pattern of breathing that will help relieve the tension and stress.

Breathing Exercises

Lie flat on your back, with knees bent. Take a couple of deep breaths. Place one hand on your abdomen just above the navel. Place the other hand just above the first. Begin breathing more slowly, evenly. Once you feel a bit more relaxed take in a very slow, deep breath through your nose, filling the lower portion of your lungs first, and slowly release through your mouth as though blowing a balloon. Repeat. Focus on your hands. When you are breathing correctly the upper hand should remain fairly still, while the lower hand rises and fall.

If you're still having trouble getting a deep breath, try this: instead of focusing on breathing in, try breathing in normally and use your energy to exhale deeply, contracting your stomach muscles to force out the last bit of air — you may find that you then automatically and naturally take in your first truly deep breath.

Another idea: post a small sign wherever you need it that says simply "breathe" as a reminder, or decide to remember to breathe every time you do a certain activity, such as every time you sit down.

The Pathway to Peace

When you need a distraction from negative thoughts and worries, take a moment to reflect on this list from Paul's letter to the Phillipians.

Focus on whatever is …

TRUE
HONORABLE
JUST
PURE
PLEASING
COMMENDABLE
EXCELLENT
WORTHY OF PRAISE

Hour By Hour

God broke our years to hours and days, that
Hour by hour
And day by day,
We might be able all along
To keep quite strong.
Should all the weight of life
Be laid across our shoulders,
And the future, rife
With woe and struggle, meet us face to face
At just one place,
We could not go;
Our feet would stop, and so
God lays a little on us every day.
And none, I believe, on all the way,
Will burdens bear so deep
Or pathways lie so steep,
But we can do, if by God's power,
We only bear the burden by the hour.

GEORGE KLINGLE

IT IS POSSIBLE TO MOVE A MOUNTAIN BY CARRYING AWAY SMALL STONES.

CHINESE PROVERB

Time

We never really notice
'til it's gone
Time passes so quickly
We let it go —
we don't hold on —
'til it's too late
God don't let another day go by
that we don't stop
to see it through.
Old times, old places, old faces
have drifted
Let us wake up to a new dawn,
our spirits uplifted.
For each and every moment we are allowed
is a gift.
A gift more precious than any other,
but one that doesn't last forever.
With every sunset
another day passes,
another moment gone —
forever.
With every sunrise
there is a new beginning,
a chance to take time…
to hold on.

C.J.D.

"From the bottom of my heart I am against those worrying cares which are taking the heart out of you. Why make God a liar in not believing His promises, when He commands us to be of good cheer and cast all our care upon Him, for He will sustain us? Do you think He throws such words to the winds? What more can the devil do than slay us? Christ died for sin once for all, but for righteousness and truth He will not die, but live and reign. Why then worry, seeing He is at the helm? He who has been our Father will also be the Father of our children. As for me (whether it proceed from God's Spirit or from stupidity, my Lord Jesus knows), I do not torment myself about such matters."

~ Martin Luther

LIVE IN THIS MOMENT

I WAS REGRETTING THE PAST AND FEARING THE FUTURE. Suddenly God was speaking. "My name is 'I Am'." I waited. God continued.

"When you live in the past, with its mistakes and regrets, it is hard. I am not there. My name is not 'I was.'

"When you live in the future, with its problems and fears, it is hard. I am not there. My name is not 'I will be.'

"When you live in this moment, it is not hard. I am here. My name is 'I Am'"

HELEN MELLICOST

Calm Waters

Let me see the future clear
Wipe away the doubt and fear
Tell me that the day is near
When I won't need to shed a tear

As I take the road less traveled
You will guide me through the battle
All the heartache all the pain
Nothing will have been in vain

Live in hope and the joy of each moment
Nothing beats a strong opponent

Calm the waters stormy sea
Shine the light so I can be
All the promise all the gifts
Everything I started with
Living life full and true
Every day wondrously new

"And the Holy Spirit shall do as He always does; He will give peace to the anxious and courage to the fearful, and a calm and gentle way to those who try to seek to do His will and to follow the command of the gospel: do not be afraid, it is I."

Rev. Benedict J. Groeschel, CFR

One Thing, Each Day

Each day finds me hopeful
looking for something special.
If I may come upon just one thing –
Then
building,
as a pool of water
forms from single drops,
each new blessing, however small,
latches on to others held before,
so that eventually I have many,
and am thankful evermore.

He who forever seeks a brighter tomorrow, and finds no happiness in today, has missed the meaning of life somewhere along the way.

SERENITY PRAYER

*God, grant me the Serenity
to accept the things I cannot change
Courage to change the things I can
and the Wisdom to know the difference.*

*Living one day at a time;
Enjoying one moment at a time;
Accepting hardship as the pathway to peace.*

*Taking, as He did, this sinful world as it is,
not as I would have it.*

*Trusting that He will make all things right
if I surrender to His Will;*

*That I may be reasonably happy in this life,
and supremely happy with Him forever
in the next. Amen*

REINHOLD NEIBUHR

AMAZING GRACE
(HYMN)

Amazing grace! How sweet the sound!
That saved a wretch like me!
I once was lost but now am found,
Was blind but now I see.

'Twas grace that taught my heart to fear,
And grace my fears relieved;
How precious did that grace appear
The hour I first believed!

Through many dangers, toils and snares
I have already come;
'Tis grace hath brought me safe thus far
And grace will lead me home.

When we've been there ten thousand years,
bright shining as the sun,
We've no less days to sing God's praise
Than when we'd first begun.

TEXT: JOHN NEWTON
TUNE: TRADITIONAL

From the author's journal...

PERSEVERE AND ENDURE

It's easy to say live in the moment.... but what about when the moment is just really lousy? There are times when we must simply persevere and endure. Those are the times in particular, when I find myself searching for the good that can possibly come of suffering or loss. The search, and endurance of heart and soul, achieved through patience, prayer, grace and the application of will, ultimately points to a couple of things: 1) The silver lining, on even the cloudiest of days, is that the pain we are forced to endure creates a unique opportunity to develop a deeper, loving connection to Christ, as we relate to His suffering. And 2), our suffering is an opportunity for others to see in us the power of God's Spirit at work, creating in us the ability to persevere with peaceful courage and often miraculous recovery. What greater good can come from suffering than our being drawn closer to Him; knowing the depth of His love and comfort, the strength of His healing power, and His will for our life and our impact on others.

REMEMBER...
NEVER QUIT BEFORE THE MIRACLE.

Great trials seem to be necessary preparation for great duties.

Find Meaning
and Purpose

Answered Prayer

I asked for strength that I might achieve;
I was made weak that I might obey.
I asked for health that I might do greater things;
I was given infirmity that I might do better things.
I asked for riches that I might be happy;
I was given poverty that I might be wise.
I asked for power
that I might have the praise of men;
I was given weakness
that I might feel the need of God.
I asked for all things that I might enjoy life;
I was given life that I might enjoy all things.
I have received nothing that I asked for,
all that I hoped for.
My prayer is answered
I am most blest.

ANONYMOUS

From the author's journal...

THIS IS YOUR OPPORTUNITY

How often have we heard someone say, after coming through a major illness or tragedy, that they were forever changed for the better. That their life has gained new meaning, they appreciate more now, their relationships are more important, and so on. Remembering this can be of great value during our own difficult times.

Seize your tragedy as an opportunity. You are not a victim of the circumstances of your life – they are a gift. This is the moment of your empowerment; the chance to find out who you can be. You may discover your true purpose in life, and find your reason for being: to give from the gifts you were born with; to share and to care and to love. Remember that the day is always darkest just before the dawn. Remember too, if you live in fear you cannot be everything you were meant to be. Trust and Live in His Peace.

THE TROUBLE WITH OPPORTUNITY IS IT ALWAYS LOOKS BETTER GOING THAN COMING.

God's Will?

"IF YOU SAY, 'WELL, IT'S A BIT CASUAL OF GOD TO ALLOW THESE THINGS TO HAPPEN IF THEY ARE NOT HIS INTENTION,' I agree there is a mystery there. It would be foolish to speak as if all the ways of God were clear... I too am often appalled at the suffering people endure... Yet I wonder if, in a sense, we are not all in the position of little children. I can imagine a child looking up at his own father who loves him, and saying to him, 'Don't you think you are rather casual to let me get hurt the way you do?'... Imagine a mass meeting of tiny toddlers...with a little toddler as chairman who addresses his fellow toddlers: 'I am sure my parents don't care. Look at my knees!'...and there they are all red and scratched. Imagine the meeting passing this resolution: 'that this meeting protests against the carelessness of parents, and demands that in future no furniture shall be made with sharp corners'... We say to God, 'Look at my frustration and sorrow and disappointment and pain! How can you be so callous, and how do you expect us to think you care?' Perhaps childhood's tragedies are to us what our tragedies are to God — not that he is callous any more than the ideal parent is, but that his perspective is different. The thought that comforts the child comforts me ... he would say, 'there is much I don't understand, but I know that my father both loves and cares.'... I am quite certain that because God is love there is nothing in his world that can be regarded as meaningless torture. There is much I cannot understand...I know that although I cannot understand the answer to my question, there is an answer, and in that I can rest content."

"The fallacy has got about that disease and suffering are the will of God... but it is the will of God only within the circumstances created by evil. Circumstantial will can be viewed from two angles — the first natural, the second spiritual. There is the physical condition [of illness and suffering], but second there is the possibility of a person making such a splendid response to that circumstance that he creates out of it a spiritual asset in the community of much more value. ... Given a spiritual awakening so glorious that the person lives in close cooperation with God, [the person] is more in line with His will. ... So many people are spiritually asleep and are not cooperating with Him at all, and so many people have, through sickness, become spiritually awakened during their illness that out of the circumstances of evil they have created and set free spiritual energies far more valuable than the spiritual apathy of the healthy person."

"I am quite sure that the battle against disease is the will of God...I like to think of our Lord standing by the bedside of the patient and working with the doctors and nurses toward the regaining of health, working on the mind and spirit of the patient as the physicians work on the body."

LESLIE D. WEATHERHEAD
EXCERPTS FROM THE WILL OF GOD

In the Midst of Suffering

I FEEL AT TIMES AS IF I COULD NEVER CEASE PRAISING GOD; COME AND REJOICE WITH ME OVER HIS GOODNESS!

I reached for Him out of my inner conflicts, and He was there to give me strength and courage. I wept in utter frustration over my troubles and He was near to help and support me.

What He has done for me He can do for you. Turn to Him; He will not turn away from you. His loving presence encompasses those who yield to Him. He is with them in the midst of their troubles and conflicts. He meets their emptiness with His abundance and shores up their weakness with His divine power.

Listen to me, I know whereof I speak. I have learned through experience that this is the way to happiness. God is ever alert to the cries of His children. He feels and bears with them their pain and problems. He is very near to those who suffer and reaches out to help those who are battered down with despair.

Even the children of God must experience affliction, but they have a loving God who will keep them and watch over them. The godless suffers in loneliness, without hope; the servant of God finds meaning and purpose even in the midst of his suffering and conflict.

Leslie Brandt, Psalms/Now, 34

May You Always Have Enough...

Happiness to keep you sweet,

Trials to keep you strong,

Sorrows to keep you human,

Hope to keep you happy,

Failure to keep you humble,

Success to make you eager,

Friends to give you comfort,

Wealth to meet your needs,

Enthusiasm to look for tomorrow,

Faith to banish depression,

Determination to make each day

Better than the day before.

Anonymous

Prayers For The Suffering

"He said to me, 'My grace is sufficient for you, for my power is made perfect in weakness.' I will all the more gladly boast of my weaknesses, that the power of Christ may rest upon me."

2 Corinthians 12:9-10

[Those that suffer] teach us that weakness is a creative part of human living, and that suffering can be embraced with no loss of dignity. Without [suffering] in your midst you might be tempted to think of health, strength, and power as the only important values to be pursued in life. But the wisdom of Christ and the power of Christ are to be seen in the weakness of those who share His sufferings. ... May God bless and comfort all who suffer. And may Jesus Christ, the Savior of the world and healer of the sick, make His light shine through human weakness as a beacon for us and for all mankind.

We thank You for the favors we receive in everything that helps us, gives us relief, and consoles us. We thank You for the medicine and the doctors, for the care of the nurses, for the circumstances of life, for those who console us and are consoled by us and accept us, and for the others. Lord grant us patience, serenity and courage; grant that we may live a joyful charity, for love of You, toward those who are suffering and towards those who, though not suffering, have not grasped the meaning of life...Amen.

Prayers of Pope John Paul II

And we rejoice in the hope of the glory of God.
But we also rejoice in our sufferings,
because we know that suffering produces perseverance;
perseverance, character; and character hope.
And hope does not disappoint us,
because God has poured out His love into our hearts
by the Holy Spirit, whom He has given us.

ROMANS 5:2-5

*Humble yourselves, therefore,
under God's mighty hand,
that he may lift you up in due time.
Cast all your anxiety on him, because he cares for you.
Be self-controlled and alert.
Your enemy the devil prowls around
like a roaring lion looking for someone to devour.
Resist him, standing firm in the faith,
because you know that your brothers around the world
are undergoing the same kind of sufferings.
And the God of all grace,
who called you to his eternal glory in Christ,
after you have suffered a little while
will make you strong, firm and steadfast.
To him be the power, for ever and ever.
Amen.*

I PETER 5:6-11

FAITH IS THE BIRD THAT KNOWS DAWN,
AND SINGS WHILE IT IS STILL DARK.

TAGORE

Divine Weaver

My life is but a weaving
between my Lord and me.
I cannot choose the colors
He works steadily
Often he weaves sorrow
and I in foolish pride
Forget that he sees the upper
and I the lower side
Not 'til the loom is silent
and the bobbins cease to fly
Shall God unroll the canvas
and explain the reason why
The dark threads are as needful
In the weavers skillful hand
As the threads of gold and silver
In the pattern He has planned

ANONYMOUS

THERE SHALL BE SHOWERS OF BLESSING
(HYMN)

There shall be showers of blessing:
This is the promise of love;
There shall be seasons refreshing,
Sent from the Savior above.

There shall be showers of blessing:
Precious reviving again;
Over the hills and the valleys,
Sound of abundance of rain.

There shall be showers of blessing:
Send them upon us, O Lord;
Grant to us now a refreshing,
Come and now honor Thy Word.

There shall be showers of blessing:
O that today they might fall,
Now as to God we're confessing,
Now as on Jesus we call!

Showers of blessing,
Showers of blessing we need:
Mercy-drops round us are falling,
But for the showers we plead.

TEXT: DANIEL W. WHITTLE
TUNE: JAMES MCGRANAHAN

A Brighter Day

I wish I could take away the pain,

I wish I could dry your tears.

But time will lift the sorrow,

and leave in its place

a brighter day.

You will find

the strength you need.

You will face each new day.

Reach out to those who care,

have the courage to share.

As you have been there for others,

they will be there for you.

C.J.D.

WHEN THE DARKEST HOUR IS PRESENT,
IT IS ALWAYS NEAR THE DAWN.

*You that live in the shelter of the Most High and lodge
under the shadow of the Almighty,[1]
Who say 'The Lord is my safe retreat, my God
in whom I trust';[2]
He will cover you with his feathers,
and you shall find safety beneath his wings;[4]
You shall not fear the hunter's trap by night
nor the arrow that flies by day,[5]
A thousand may fall at your side, ten thousand
close at hand,
but you it shall not touch; His truth will be your shield[7]
For you the Lord is a safe retreat;
you have made the Most High your refuge.[9]
No disaster shall befall you[10]
For he has charged his angels to guard you wherever you go[11]
Because his love is set on me, I will deliver him;
I will lift him beyond danger, for he knows me
by my name.[14]
When he calls upon me, I will answer;
I will be with him in time of trouble;
I will rescue him and bring him to honor.[15]
I will satisfy him with long life to enjoy the fullness
of my salvation.[16]*

PSALM 91

On The Other Side

On the other side of fear is opportunity
On the other side of pain is peace
On the other side of doubt is faith
On the other side of worry is trust
On the other side of rain is a rainbow
On the other side of cloudy skies is sunshine
On the other side of harsh words is music
On the other side of tension is release
On the other side of humbled is humor
On the other side of injustice is reward
On the other side of helplessness is hope
On the other side of forsaken is uplifted
On the other side of emptiness is fulfillment
On the other side of loneliness is a friend
On the other side of temptation is a conscience
On the other side of weakness is strength
On the other side of disaster is recovery
On the other side of why is enlightenment
On the other side of lost is the way
On the other side of tears is laughter
On the other side of darkness is light
On the better side of all is love

C.J.D.

Prayer of Thanksgiving

O Loving Father, whose healing power has touched my life in its hours of pain and distress, and has now set it upon the road to recovery, I thank you for that strength which has placed me upon this road. You did sustain me in my weakness and support me in my pain. As you then gave me energy to endure, so now, O Lord, give me patience to persist to the completion of my recovery. Help my continuing thanksgiving to be in finding your will for my life and in following it from this day and evermore. Through Jesus Christ, our Lord. Amen.

Robert Rasche

Let our thanks be in service to You.

Receive the Gifts of Peace, Guidance, and Healing

I GENTLY CLOSE THE DOOR TO ALL OUTER DISTRACTIONS AND MEET WITH GOD IN THE SILENCE OF MY BEING. In the silence, I have a renewed awareness of God and God's gifts for me.

In silence I receive the gift of peace and accept it into my life now. Peace is the cup I hold forth to be filled with all the blessings I am prepared to receive.

In silence, I receive the gift of guidance. It fills me, surrounds me, and lights my way. I go forth, living and walking in God's wonderful revealing light.

In silence, I receive the gift of healing. Healing springs forth from deep within me now. I am whole, well, and strong.

In silence, I claim God's gifts to me and say, "Thank You, God, for peace, guidance, and healing."

DAILY WORD, EDITED BY COLLEEN ZUCK

Is anybody happier because you passed his way?

Does anyone remember that you spoke to him today?

Can you say tonight, in parting with the day

that's slipping fast,

That you helped a single brother

of the many that you passed?

Is a single heart rejoicing over what you did or said?

Does the man whose hopes were fading

now with courage look ahead?

Did you leave a trail of kindness,

or a scar of discontent?

As you close your eyes in slumber,

do you think that God will say,

"You have earned one more tomorrow

by the work you did today"?

~ Anonymous

From the author's journal...

MAKING A DIFFERENCE

Once you come through a crisis it can be more difficult than you might expect to let go of the pain and put the experience behind you. I know I was surprised at my inability to simply feel the relief of "it's over" — it was as though something more was necessary. What I discovered was the need to transfer the newly gained awareness, strength, and energy I had been exerting throughout the battle to a new goal: making the world a better place. Sounds lofty, but where would we be without all of the people who have made a difference? The pain I endured left me with much more compassion and a profound sense of the need to help my fellow human beings, and I know I'm not alone in this. You can turn your pain into the basis for giving comfort and help to others. Remember, God has a bigger plan for you than you can possibly dream for yourself. If you can find God's will for your life your efforts toward that end will be successful. Your work will be meaningful and you will be happy in it.

So…
Go forward now.

Journal

INDEX

FORWARD

| Poem | Emily Dickenson | 2 |
| Journal | First Call For Help | 3 |

KNOW THAT YOU ARE NOT ALONE

Note	Turning To God	6
Journal	The Peace of His Presence	7
Hymn	On Eagle's Wings	8
Scripture	Isaiah 40:28-31	9
Poem	Footprints	10
Psalm	Psalm 121	11
Journal	It's In Your Heart	12
Scripture	Ephesians 4:16-19	13
Note	My Heart Is Secured	14
Psalm	Psalm 71	16
Poem	With His Help I Am	17

GAIN THE STRENGTH TO HEAL

Exercise	Healing Energy	20
Journal	Emotional and Physical Healing	21
Exercise	An Exercise of Faith	22
Poem	The First Ray	24
Exercise	Visualization	25
Exercise	Quietness	26
Exercise	Smile	27
Exercise	Feed Your Soul	28
Journal	Pray	30
Hymn	What A Friend We Have In Jesus	31
Prayer	Daily Prayers	32
Prayer	Prayers for Healing and Surgery	33
Psalm	Psalm 103:1-5	34
Journal	Read the Bible	35
Exercise	Become Centered and Balanced	36
Exercise	Reflect In The Beauty Of Nature	37

Don't Worry About Tomorrow

Poem	Ralph Waldo Emerson	40
Journal	Fasting From Worry	41
Prayer	Prayer On Worry	42
Note	Yesterday, Today & Tomorrow	43
Scripture	Matthew 6:25-34	44
Exercise	Don't Worry, Breathe Deeply	45
Exercise	The Pathway to Peace	46
Poem	Hour by Hour	47
Poem	Time	48
Note	Martin Luther	49
Note	Live In This Moment	50
Poem	Calm Waters	51
Poem	One Thing, Each Day	52
Prayer	Serenity Prayer	53
Hymn	Amazing Grace	54
Journal	Persevere and Endure	55

Find Meaning And Purpose

Poem	Answered Prayer	58
Journal	This Is Your Opportunity	59
Note	God's Will?	60
Note	In The Midst of Suffering	62
Poem	May You Always Have Enough	63
Prayer	Prayers For The Suffering	64
Scripture	Romans 5:2-5	65
Scripture	I Peter 5:6-11	66
Poem	Divine Weaver	67
Hymn	There Shall Be Showers of Blessing	68
Poem	A Brighter Day	69
Psalm	Psalm 91	70
Poem	On The Other Side	71
Prayer	Prayer of Thanksgiving	72
Note	Receive the Gifts of Peace, Guidance, and Healing	73
Note	Is Anybody Happier...	74
Journal	Making A Difference	75

Help is Here!
A Spiritual Survival Manual *for* Times of Crisis.

Cop If you know someone else that is in need of   website at
     this book, please contact Molly DeLong:
     Phone: 612-616-0690
     Email: vanced@mn.rr.com   to:

ORDER FORM

NAME _____

ADDRESS _____

CITY _____

STATE _____ ZIP _____

PHONE _____

EMAIL _____

PAYMENT   (circle one)   CHECK   CREDIT CARD

CREDIT CARD NUMBER _____

NAME ON CARD _____

EXP DATE _____

ORDER QTY _____ X $11.95 + Shipping* = $_____

*$3.00 shipping first item, $1.50 each add'l item.
Mail to: Excelsior House Publishing — *Help is Here!*
             PO Box 623, Excelsior, MN 55331

EXCELSIOR HOUSE PUBLISHING, Inc.